Animals, Work, and the Promise of Interspecies Solidarity

Animals, Work, and the Promise of Interspecies Solidarity

Kendra Coulter

First published 2016 by
PALGRAVE MACMILLAN

The author has asserted their right to be identified as the author of this work in accordance with the Copyright, Designs and Patents Act 1988.

Palgrave Macmillan in the UK is an imprint of Macmillan Publishers Limited, registered in England, company number 785998, of Houndmills, Basingstoke, Hampshire, RG21 6XS.

Palgrave Macmillan in the US is a division of Nature America, Inc., One New York Plaza, Suite 4500, New York, NY 10004-1562.

Palgrave Macmillan is the global academic imprint of the above companies and has companies and representatives throughout the world.

Hardback ISBN: 978–1–137–55879–4
Paperback ISBN: 978–1–349–71892–4
E-PUB ISBN: 978–1–137–55881–7
E-PDF ISBN: 978–1–137–55880–0
DOI: 10.1057/9781137558800

Distribution in the UK, Europe and the rest of the world is by Palgrave Macmillan®, a division of Macmillan Publishers Limited, registered in England, company number 785998, of Houndmills, Basingstoke, Hampshire RG21 6XS.

Library of Congress Cataloging-in-Publication Data

Coulter, Kendra, 1979–
    Animals, work, and the promise of interspecies solidarity / Kendra Coulter.
        pages cm
    Includes bibliographical references and index.
    ISBN 978–1–137–55879–4 (alk. paper)
        1. Animal rights. 2. Human – animal relationships – Economic aspects.
    3. Working animals. 4. Labor. 5. Precarious employment. 6. Feminist theory.
    I. Title.
HV4708.C68 2015
179'.3—dc23   2015020694

A catalogue record for the book is available from the British Library.

Contents

Figures

Acknowledgments

This project is truly a labor of love, and it reflects both necessity and possibility. It is inspired by sentient beings across species, by their courage and altruism, by their fear, sadness, and joy. The seeds of this book were planted when my mother, Rebecca Priegert Coulter, began teaching me about animals and how to respect them, on our acreage, even before I could walk. I am sincerely grateful for her foundational and ongoing encouragement and generosity. Throughout my life I have learned from others: educators, writers, researchers, advocates, activists, caregivers, organizers, artists, and workers of all kinds. The work done to improve lives, to grow, to heal, to teach, to create, to expose, to challenge, to resist, to support, to magnify, to understand, and to foster change matters.

Academic mentors and colleagues have helped fertilize this project in different ways. Arja Vainio-Mattila nurtured my early interest in gorillas and social justice. More recently, Larry Savage, the current director of the Centre for Labour Studies at Brock University, encouraged me to create our unique "Animals at Work" course, which has helped enshrine human-animal labor as an area of study, and permitted me to teach, engage with, and inspire a diversity of students. I am particularly grateful to Will Kymlicka, Susan Nance, Keri Cronin, Nik Taylor, and Janet Miller, who read and offered helpful feedback on sections of the manuscript. Lauren Corman, Amy Fitzgerald, Simon Black, Stephanie Ross, Stefan Dolgert, Margo DeMello, Jocelyne Porcher, Hilary Cunningham, Meg Luxton, Susanna Hedenborg, Emma Varley, Tom Dunk, Jonathan Clark, Neil Brooks, and Donna Baines have also supported this project and/or helped my thinking in different ways. Kirsten Francescone and Anelyse Weiler provided excellent research assistance. I am also grateful for good friends, including Rick Telfer, Violet Chavez Osorio, and Jorge Emilio Roses Labrada.

Brock University has facilitated the research that informs this book, including through the Jobs and Justice Research Unit, the Social Justice Research Institute, and the Council for Research in the Social Sciences. I appreciate the work of our administrative assistant Elizabeth Wasylowich, and staff across campus, particularly in the Office of Research Services, and in Marketing and Communications. I also want to acknowledge the people of Palgrave Macmillan, especially my editor, Stacy Noto, for her enthusiastic support of this book and area of study, and Marcus Ballenger for his editorial assistance work.

My husband, John Drew, always offers meaningful intellectual insights, continuous encouragement, and genuine, uplifting, loving partnership. I am deeply and infinitely grateful to and for him, and for our multispecies family. I have been honored to spend time with animals of all kinds who have given and shared more than I could ever adequately summarize: Buster, Sunny, Kozzie, Magic, Chlöe, Macey, Trooper, Quinn, Aquila, Panda, Dierks, Phoenix, Amigo, Sarge, Rusty, Paz, Hazy, Sophie, Champie, Jello, among countless others. This book is dedicated to you, to the animals I will never meet but whose lives matter just as much, and to the people who know—and who will know—that our species must do better.

In interspecies solidarity.

Introducing Animal Work

Interest in human-animal relations is growing across contexts and cultures. Certain people and communities have always recognized that humans are but one of many species that share this planet. Others are now more fully recognizing this fact and reflecting on what it means for individuals, whole species, and our shared futures. Yet despite this burgeoning curiosity and inquiry, the diverse and complex realities of human-animal work relations remain relatively underexamined and are not well theorized. They may also be seen as a source of tension, immobility, and division. Labor advocates may reject the idea that animals do work or see the prospect of greater concern for animals as disconnected from human workers' struggles, as frivolous, or as a threat to jobs, no matter how precarious, unpleasant, or damaging this work may be. Alternatively, among those passionate about animals, human workers and labor organizations may be criticized as merely part of the problem, and/or the fact that most people need jobs in order to survive may simply be ignored and avoided.

Yet, work has a profound impact on billions of human and animal lives, and consequently warrants more serious consideration and analysis. Many human and animal work-lives are entangled or even interdependent. Work influences income and material security, bodily health and mental well-being, knowledge of ourselves and others, and perceptions of what is desirable and possible. It is true that workplaces are where the most widespread and extreme examples of violence against animals occur. But spaces of work are also sites of compassion, devotion, learning, resistance, and possibility. In multispecies relationships, we can see the best in people, how much and how many humans benefit from animals, and examples of how people could act more ethically. By thoughtfully exploring animals and work, and moving beyond our intellectual cages, we gain a deeper and fuller understanding of labor and of the true range of the world's workers. Moreover, we can glean meaningful ideas about how to build more inclusive, ethical, holistic, and inspiring workplaces and societies. Accordingly, in this book, I unsettle and expand conventional understandings of both work and animals. I propose a multifaceted and contextualized approach to understanding work involving animals, one that is genuinely multispecies and that takes both human and animal well-being seriously. In other words, I argue that we ought to not only think deeply and differently about animals and work, but also to reflect on the promise of interspecies solidarity.

The concept of animal work serves as both an umbrella term and a springboard for examining labor involving animals and the many intersections of animals and work. The term "animal work" was used briefly by Arnold Arluke and Clinton R. Sanders (1996) to mean work with animals, and it is employed by Jocelyne Porcher (2016) to mean work done by animals. It is also used very differently by psychological and medical researchers to refer to research and/ or testing on animals. Nik Taylor (2013) has framed animal shelter and welfare advocacy efforts as work with/for animals. I expand on these usages. I use the concept of animal work as an organizing framework to highlight and think about the work done *with, by,* and *for* animals. To more fully develop this concept and encourage thorough understanding, the subcategories will be further elucidated as well as combined.

I first examine the work done with/for animals. In chapter 1, I illuminate the daily work and labor processes that are involved when people work with and/or for animals across sectors, and emphasize both the material and experiential dimensions of work. Next, I focus on the work done by animals themselves. In chapter 2, I examine and unpack the breadth of animals' work to broaden the labor lens and posit theoretical concepts and frameworks that foster deeper understanding of these processes and of animals' own experiences. I analyze examples that are widely and clearly identified as work, but also propose an expanded lens for seeing and understanding a fuller range of animals' labor. This necessitates consideration of work that is harmful, work that is positive, as well as dynamics that do not tidily fit into a single category. The third chapter again spotlights work done with/for animals, but now approached as a broader, advocacy-oriented sphere; that is, as political labor. I first examine two central vehicles: animal advocacy/protection work and labor unions, and highlight the work of advocacy itself. I then consider the role of the public sector, as both a space of policy-making and as an employment sphere.

This book encourages deeper understanding of both work and animals, and proposes a more expansive scholarly approach to both areas. My concerns are not only theoretical, however; I also seek to foster political change. As a result, drawing in particular on feminist political economy, I conclude by reflecting on what conceptual and practical lessons can be learned about animal work overall, and about the challenges and possibilities of empathy and connectivity across species. This commitment to seeing and encouraging connections takes up the call by Claire Jean Kim (2015) for "multi-optic vision," a decidedly intersectional approach to multispecies politics. I propose the concept of interspecies solidarity as an idea, a goal, a process, an ethical commitment, and a political project that can help foster better conditions for animals, improve people's work lives, and interweave human and animal well-being. I introduce the idea of sustaining and creating humane jobs—jobs that are good for people and for animals—and emphasize the need for a longer and larger conversation about job creation, job quality, and dignity across species. Thus I take seriously the question Lynda Birke (2009, 1) asks of those studying human-animal relations: "what's in it for the animals?" I also extend this question to human workers; my scholarship recognizes the importance of working-class and poor people's experiences and lives. I seek to generate knowledge and possibilities that can make a meaningful difference not only for work, but for work-lives, both human and nonhuman.

Indeed, this is a book about both people and animals. As such, it highlights work that is multispecies, or that involves multiple species. Alan Smart (2014), among others, has pointed out that multispecies inquiry can include plants and other organisms. Here I focus on sentient beings and various species of animals. When multiple species are present, they interact, thus interspecies work, or the relationships between species, is also pertinent. How members of different species relate, whether through physical touch and their senses, active collaboration, directives and directing, domination, various kinds of communication, relationship-building, and emotional connectivity or detachment, all shape the realities and experiences of work.

Before delving into the material, a brief comment on language is in order. Humans are also animals; humans evolved on earth. As a result, some writers use the term "nonhuman animals." For the sake of linguistic simplicity, and to avoid continuously referring to other species by what they are not, I use the terms "people" or "humans," and "animals." I also use "he," "she," and "who" to refer to animals, rather than "it" and "that," to reflect the fact that animals are sentient beings who perceive and feel, not only pain and suffering, but also sadness, fear, and joy.

Work and labor are, of course, also essential concepts in this book. Yet there is no consensus about the precise definitions of work and/or labor among researchers. For example, some scholars see work as the process of working and labor as a term capturing a larger set of political interests such as organized labor or a broader working-class alliance. Others see work as a set of tasks and conceptualize labor as a social process. Yet others have different ways of understanding the connections and distinctions between these terms. In this book I use both terms. When seeking to use terms in specific ways, I provide additional details, adjectives, and/or corollary concepts to indicate my precise intention.

Working Creatures Great and Small

People have always lived around other species. As Susan Nance (2013a, 7) points out, "There has never been any purely human space in world history." Not surprisingly, people have always worked around and with other species, whether it was explicitly called work, was seen as mere subsistence, or was deemed part of "development," conquest, and/or imperialism. For the past few million years—at least 90 percent of human history—evolutionary ancestors and earlier groups of the current human species, *Homo sapiens*, were foragers/hunter-gatherers. Until about 10,000 years ago, most groups of humans subsisted primarily because they (and especially women) collected plants, roots, nuts, seeds, berries, and so forth. In most communities, meat, obtained through hunting other animals, fishing, and/or scavenging, comprised a small portion of the diet. Some animals deliberately avoided direct contact with people, despite sharing terrain, and there were certainly animals our ancestors avoided, as well. Yet small, nomadic bands of people moved often, and the presence of various other animals across natural spaces was continuous; the world was teeming with animal life.

Humans' earliest relationships with one species continue to this day. Anthropological and biological research has found that dogs and people began living in close proximity and forming bonds 10,000 years ago, with some data suggesting it was as far back as 30,000 years ago (Larson et al. 2012; McCabe 2010). Most notably, there is debate among researchers about the precise processes of domestication and human-canine relationship-building. Researchers differ in their interpretations of whether humans domesticated dogs, dogs actively chose to live by and with people, or if it was a bit of both. Some researchers argue that dogs were, in fact, essential to the survival and thriving of early humans, in contrast to Neanderthals who declined and then died out (Shipman 2012). Without question, work was central to these early relationships, as dogs shaped human's hunting, provided some security and protection, and offered companionship—and vice versa.

Some foraging groups settled and became subsistence farmers/horticulturalists, expanding the number of species with whom people interacted and changing the nature of those relationships. Certain of the small farming communities grew in size, and later expanded into the first state societies. Such changes again intensified and significantly altered the natural environment, including animals' lives. The global portrait of human evolution includes great diversity as well as common features well beyond the scope of this discussion, but the universal involvement of animals is noteworthy. Animals were central to the formation of every human society and mode of production on the planet. Different wild animal species, including sheep, cats, goats, pigs, cows, chickens, donkeys, ducks, camels, llamas, and alpacas were all domesticated in the last 13,000 years, as humans sought to corral and use animals' bodies, minds, and abilities (Zeder 2008).

At the same time, by altering animals' existences and changing the natural environment everywhere we went, unintended human-animal interactions began or were exacerbated. The number and frequency of mosquitoes and mosquito-borne illnesses grew alongside the pools of water and piles of excrement that accompanied agriculture, for example. Human settlements attracted more rodents and scavengers. Close living and working with animals meant an increased intermingling of living and dead bodies, bodily fluids, and microorganisms (Swabe 1999). People also began shaping not only animals' lives, but also their biological futures and genetic makeup through selective breeding and other practices that continue even today (Ritvo 2010). Overall, the number of species and individual animals whose futures were wedded to human beings is astounding. From the smallest insects, birds, and mice, to the massive whales and elephants, no animals or species were unaffected by humans and their work.

Researchers are largely in agreement that tool usage, language, and the domestication of animals were the most influential drivers of human biological and social evolution. Without question, animals' work also propelled human societies. Horses have moved people and goods, provided power for the development of agriculture, manufacturing, communications systems, and infrastructure, and been enlisted for warfare, conquest, sport, leisure, and companionship. Historians and anthropologists alike argue that the benefits humans gained from horses are immeasurable and that horses were essential to the ascendancy of human civilizations (Chamberlin 2006; Greene 2008; McShane and Tarr 2007). I have argued that "The history and contemporary state of human-horse interactions are underscored by two main related factors: horses do work of various kinds for people, and people garner material, social and/or personal gain from that work" (Coulter 2014a, 148). This statement can be extended beyond equines, to aptly characterize animals' work more broadly.

At the same time, the history of human-equine relationships also illustrates how people's connections to animals were and are not only material and utilitarian, but also symbolic, emotional, and personal. Cave drawings and engravings of horses date as far back as 3,500 years ago, and the written and oral histories of Aboriginal peoples tell of meaningful and evocative relationships with individual horses and the species as a whole (Chamberlin 2006; Horse Capture and Her Many Horses 2006; Lawrence 1985). Pat Shipman (2011) argues that people's capacity to care about and for animals was also pivotal for human evolution and the development of societies.

Whether and to what degree animals have benefited from their relationships with people is fiercely debated, however, and researchers hold vastly different views. Some argue that mutual reciprocity characterizes the history of human-animal relationships, and that animals not only chose domestication, but also benefit from their connections to humans (Budiansky 1999). In contrast, David A. Nibert (2013, 2) defines the process not as domestication, but as "domesecration," a process that, in fact, "*undermined* the development of a just and peaceful world" through the institutionalization of domination, not only of animals, but also of devalued humans and particularly indigenous peoples. Because I was trained as an anthropologist, I recognize the importance of seeing patterns and systems, but also the limitations of broad claims that purport to reflect the entire, global picture. Local differences and specifics complicate most attempts to speak universally. Moreover, these local specificities matter and provide a more complete picture. There is abundant evidence of domesecration and domination (Ingold 1994), but there are also examples of human-animal relationships characterized by respect, love, and/or reciprocity. The denial of either means a failure to capture the complexities of human-animal relationships. Recognizing that there are positive human-animal relationships also does not detract from the crucial work of identifying and challenging forms of oppression. In fact, in my view, it is by understanding both the areas of harm and the dynamics of hope that we can gain the most thoughtful, thorough, and helpful insights about how to reduce suffering, improve lives, and foster humane action. What should be changed and nurtured, both become clear, particularly if we pay close, critical, and thoughtful attention to spaces and relations of work. Thus, in this book I consider work in a range of contexts, urge nuanced, evidence-based understanding, and purposefully eschew generalized, totalizing statements.

These divergent intellectual perspectives reflect not only interpretations of the evolutionary and historical record, but also some of the broader political and theoretical weather systems within which contemporary human-animal studies and this book are situated. Because this book's focus is unusual and politically engaged and its emphases broad, my analysis intersects and engages with a diverse cross-section of literatures, particularly from labor studies, human-animal and critical animal studies, feminist political economy, and certain strands of ecofeminism. I employ a gendered and intersectional lens throughout the book. I also enlist ideas from the sociology of work, anthropology, animal welfare and rights theories, animal and feminist ethics, environmental history, animal-focused social histories, cognitive ethology, political ecology, and critical geography. Accordingly, I will identify certain broad intellectual trends below, but more thoroughly contextualize the discussion within and draw from pertinent areas of inquiry throughout the book.

Navigating the Multispecies Intellectual Landscape

Debates, differences, and disagreements shape the interspecies and multispecies intellectual terrain within and across cultures. Presenting a full picture of the breadth of interspecies ideas, symbols, and interactive relationships cross-culturally is well-beyond the scope of this discussion. Anthropological research, in particular, illuminates the vast array of ways that people have thought and do think about animals; it also reveals how researchers have differently and similarly interpreted and analyzed these data. Without question, indigenous peoples across the globe have a long intellectual history of reflecting on the roles and meanings of different species. These communities tend to envision a more holistic and less hierarchical relationship with nonhumans, overall. Yet, while there is a strong sense of connectivity with the past in many indigenous cultures, no peoples' ideas about animals and human-animal relationships are fixed, unchanging entities. Notably, indigenous epistemologies or worldviews are also diverse and context-specific, and they are continuously shaped by the cultures, environments, historical processes, and agency of the social actors involved (Brightman 1993; Harrod 2000; Kailo 2008; M'Closkey 2002; McHugh 2013; Morris 2000; Pomedli 2014).

Much Western/colonial intellectual history, particularly post-Enlightenment, involves the construction and reproduction of dichotomies: nature-culture, emotion-reason, woman-man, among others (Adams 2007; Descola 2013; Ingold 1994). In addition to being presented as poles or binaries, these couplets are conceptualized hierarchically, with culture, reason, and man generally framed as superior. All three sets of dichotomies are relevant to this book, but the nature-culture binary is the most pertinent here. Colonial societies have consistently sought to demarcate themselves as different from and above "nature," which includes animals and the natural world or environment. "Culture" has been seen as the domain of humans, as a unique testament to people's intelligence, morality, and superiority. This continued despite the fact that humans are biological beings who get sick and hurt and die, like every other creature on earth. It is, of course, true that humans developed impressive skills such as writing, as well as fields like art and science. Moreover, all social, political, and economic systems are socially constructed, not naturally determined. Across space and time, however, too many humans have often used their intellectual abilities and technologies to cause harm to other people and animals.

At the same time, the nature-culture dynamic was never entirely dichotomous or consistent. While humans do seek to differentiate themselves from animals, people also use biological explanations for their behavior. Competition, infidelity, greed, poverty, war, and many other attributes and practices have been credited to human "nature" and described as "natural," not only by certain scientists, but also by regular people. People also enlist countless biological and animalistic metaphors, images, and descriptors as compliments, insults, and everything in between. It is a rather glaring contradiction that the one species that uses its difference and alleged exceptionalism as justification for countless harmful practices simultaneously evokes its nature and biology as the reason and justification for yet other harmful practices.

Indeed, in the last decades of the twentieth century, countless critical, cross-cultural, and feminist writers have challenged the rigidity, universality, and hierarchy of these dichotomies in many ways. With respect to nature and culture, this includes recognizing that animals and nature are present in and fundamentally interwoven with social relations and structures in countless ways. Nature and representatives thereof are present in all spaces of culture, and essential to all economic activity. Moreover, animals have been found to have their own culture. In fact, some species have been found to have difference cultures (Van Schaik et al. 2003; Whiten et al. 1999).

Crucially, these insights should not translate into an abandonment of social constructivism or the embracing of biological reductionism or essentialism, however. As Val Plumwood (2002, 201) puts it, "although we need to affirm continuity with nature to counter our historical denials, doing so does not require any simple assumption of identity." An imprecise or decontextualized emphasis on humans as animals can contribute to or reproduce incorrect naturalistic ideas and explanations for people's behaviors which stem from cultural patterns, processes of socialization, and political and ethical choices. Animal advocates who seek to completely dissolve the human-animal boundary can find themselves in tricky territory when arguing against the consumption of nonhuman animals—because omnivores in the natural world do eat other animals sometimes, and humans are biologically omnivorous. Proponents of animal production and consumption can accuse animal advocates of wanting their cake and eating it too: humans are animals, but humans should still not consume animals even though that is done in nature. The response to this charge is usually that while humans have historically consumed some animals, they do not need to do so, and particularly not today (in the large majority of places) given the evolution of food production and distribution systems. In other words, people, as both biological and social beings, can make different ethical choices to improve and protect the lives of others. For me, this latter fact is the most important. We can and should make socially determined political decisions about how we act and treat others within and across species, whether they are similar to or different from us.

Intellectually, we now recognize the *intersections* of nature and culture and the overlapping space between the two spheres. Terms like natureculture, naturalcultural, socionatural, and the natureculture nexus are used to recognize post-dichotomy approaches and the border and contact zones (see, e.g., Haraway 1989, 1991). This approach does not mean that inequities, whether based on gender, species, class, race, and/or other factors are denied or naturalized. Rather, we can identify and understand the socioeconomic relations, hierarchies, and structures that exist and are sustained or changed in these intersections, and the need to "attend" to nature and animals in social analyses (Gaard 2011; Gruen 2009). Broad-brush concepts like "animals" are useful but also warrant further nuance. The term "animals" represents a very broad cross-section of groups and individuals, and sentient nonhuman beings (like humans) have similarities as well as differences. This is true across and within species. At the same time, human societies and individual people the world over see and treat animals in divergent and often contradictory ways (Francione 2008; Herzog 2010). This includes purporting to love animals and cherishing some of them, while condoning the exploitation and/or killing of many others for food, clothing, sport, and/or research. Even allegedly loved species (e.g., dogs, cats, and horses in Canada and the United States) face very uneven treatment, as some individuals are celebrated, while others are devalued, abused, and/or killed.

Despite the growing body of research on the entanglements of human and nonhuman lives, an exclusionary focus on people persists in many academic quarters, including in labor studies and in the study of work across disciplines. A large majority of labor researchers fail to see that humans are but one of many species in any given space and community, that many humans work with animals, that humans depend upon the broader ecological web for subsistence and survival, and that human, animal, and environmental well-being are inextricably connected. Undoubtedly, the work done by animals themselves is particularly underexamined. Far too often, analysis stops at humans, and so does empathy. At the same time, political economy, and labor questions in particular, remain underexplored in the interdisciplinary field of animal studies. Human-animal relations are often studied without giving due attention to the structural and contextual factors shaping both human and nonhuman lives (Twine 2013).

In contrast, the lens of animal work illustrates the interconnectedness of nature and culture, and the importance of context. This includes recognizing that structural factors shape and constrain both people and animals, and that both have agency. It means seeing that people and animals interact in spaces and relationships of work, that elements and beings from nature are shaped in mixed species spaces, that animal workers adapt to human demands and needs, and that animals shape multispecies worksites. Accordingly, this book is also situated in what I call the nature-labor nexus, or in the intersections of nature and labor (Coulter 2014a). This conceptualization can be applied more broadly to think about the many spaces of work involving nature. Animal work is a fruitful and important interface within which tens of billions of lives are situated at any given moment. By recognizing the nature-labor nexus and animal work, we are encouraged to think about entire species *and* individual animals. We recognize but move beyond boundaries like "wild" and "domesticated," and "the environment" versus "animals," in favor of a more integrated approach. By using animal work as a conceptual engine, we also complicate other binaries like "urban" and "rural," as humans and animals live, work, move, are moved, and are killed across spaces in a more fluid dynamic (Peggs 2012). As Raymond Williams (1975, 296–7) writes, "Most obviously since the Industrial Revolution, but in my view also since the beginning of the capitalist agrarian mode of production, our powerful images of country and city have been ways of responding to a whole social development. This is why, in the end, we must not limit ourselves to their contrast but go on to see their interrelations and through these the real shape of the underlying crisis." We are also prompted to think about the relationships between cultural and material processes, between what is thought and what is done. Peter Dickens (1996, 107) stresses "that the separation of human beings from nature is not simply the result of people having the wrong ideas about nature. Loss of biodiversity, the thinning of the ozone layer and so on, are not occurring simply because we have the wrong ideas. Rather, they are results of how human societies have worked on nature and how such work has led to, and been assisted by, wrong ideas."

Indeed, understanding and action are connected, and, often, are mutually reinforcing. At the same time, how we experience the world shapes and can change how we understand it. Equally important is that the reverse can also be true. Knowledge is political. Ideas and theoretical frameworks are not developed in a vacuum but rather in specific contexts and historical moments. Ideas are generated by thinkers, and those thinkers are shaped by the social, political, economic, and cultural contexts within which they live and work. In particular contexts, certain ideas are more easily generated, while others are obfuscated, resisted, marginalized, and/or mocked. Individuals will also experience the same contexts differently depending on factors like their gender, race, ethnicity, age, nationality, and so forth and because of their individual personalities, politics, and choices. Moreover, the generation of ideas is also a collective endeavour, and every writer extends, contests, adjusts, resists, tempers, nuances, and/or creates knowledge, insights, and theory based on engagement with the work of others. All academic work is political in some way, whether through what is asked, what is highlighted, what is avoided, how it is funded, and/or who it serves. Both emphases and silences speak volumes about priorities, privilege, and marginalization.

I do not pretend to be a dispassionate writer; a political and ethical commitment to ending suffering and improving lives inspires my work. This ethic does not replace the process of data collection and analysis, however. It also does not mean I pursue or propose a singular or reductionist path, or pick and choose data to prove a particular point. In contrast, my intellectual commitment is to empirically based research, thoughtful inquiry, and critical reflection on multiple possible paths. I am intellectually and emotionally committed to improving people's and animals' lives, thus I harness those feelings into intellectual rigor and multifaceted analysis. As such, this book exemplifies what Nik Taylor and Richard Twine (2014) identify as engaged theory, a hallmark of the growing and diversifying field of critical animal studies. Critical Animal Studies (CAS) involves a particular approach to human and animal liberation, one that generally advances an abolitionist approach to animals (i.e., no human ownership or use of animals). A broader small c critical animal studies literature and community is also growing within which a broader set of political positions on animals and social justice are advanced. Both clusters of critical animal scholarship share a commitment to engaged theory. Notably, such a commitment is also an established and central part of labor studies. Both critical animal studies and labor studies see a concern for social justice as an important, laudable, and necessary component of scholarship and scholars' work, yet only the former explicitly and deliberately extends such principles beyond humans.

Overall, I engage in what I call the scholarship of possibility. This means I not only seek to build nuanced understanding of the way things really are, but I also propose changes and alternatives that help foster progressive change. In this book, I argue for new ways of thinking, first and foremost, as well as conclude by offering political ideas for thoughtfully moving forward. This includes easily achievable steps and broader, transformative ideas. Notably, I do not advance a single proposal or purport to have all the answers. In fact, I see the promotion of simplistic or totalizing explanations as unhelpful, and this book refuses to simply reproduce conceptual and political frames that either celebrate or condemn. The realities are too complex and the issues too significant for mere sloganeering, rhetoric, or ideological rigidity. There are persuasive ethical, economic, environmental, and social reasons why we need to take animal work seriously. My approach to the challenge is to share analysis and different ideas that are informed by evidence and propelled by hope. This book is driven by a labor studies commitment to taking work and workers seriously; a gendered and intersectional awareness of interlocking oppressions and how the political is personal; an anthropological ethic of multifaceted inquiry and dialogue; and by a transdisciplinary commitment to breadth. I also bring many years of experience with/in the country. I understand and respect the complexities and heterogeneity of rural and urban interspecies communities alike, and fully recognize the need for sustainable rural livelihoods.

At the same time, you are not an apolitical and ahistorical reader. You too are shaped by your context and position therein. You approach this book with ideas about and experience with people and animals, with political beliefs and assumptions, and with varying degrees of curiosity, cynicism, and open-mindedness. Thus, it is important to recognize these positions, to be self-reflexive, and to build and gain knowledge through greater understanding. In this book I present and analyze data, propose new approaches, and offer food for thought—and action.

Conceptually and methodologically, the research process that informs this book was inductive, which means the theoretical contributions offered and the framework presented here stem from evidence. I did not begin this research seeking to prove a specific intellectual or political point about animal work, nor was I aware of what this book's emphases and arguments would be. I did not impose a pre-prescribed theory or test a set hypothesis. Rather, because of my interest in human and animal well-being, I began researching and reflecting on the broad themes of animals and work, and have developed the concepts of animal work and interspecies solidarity, and the other proposals contained in this book, based on data and by building on both intellectual and political context.

The data has been collected from various sources and through a number of methods. I combine primary data sources including legal documents, policies, reports, statistics, and case studies, with the analyses of other researchers and social theory. Workers' first-hand experiences and perspectives, shared through public statements, interviews, and participant-observation are also considered. Although I was trained as an anthropologist, this book is not an ethnography, a core data collection and presentation strategy used by sociocultural anthropologists who undertake extended participant-observation in a specific local community (or set of communities) and then share their insights and analyses. This book does, however, incorporate findings collected through three years of participant-observation research on human-horse relationships in Ontario, Canada, an example of a growing scholarly interest in multispecies ethnography (Hamilton and Taylor 2012: Kirksey and Helmrich 2010; Maurstad, Davis, and Cowles 2013; Smart 2014). The ethnographic insights thus contribute to the overall methodological mix. Through these different data collection strategies, I have sought to study and, in turn, portray a holistic picture.

A focus on animals does not only form a core plank of my conceptual approach to animal work, but was also integral to the research process. Time was specifically and thoughtfully allocated for observing animals, particularly horses and dogs, and for seeking to understand their experiences. In this task, direct observation was supplemented by reading of cognitive ethology, behavioral science, and animal welfare science to augment my ability to understand animals on their own terms and correctly interpret their behavior, a strategy suggested and used by Susan Nance (2013a). Relationship-building and humility are essential, as well; as A. A. Milne put it in *Winnie the Pooh*, "some people talk to animals. Not many listen though. That's the problem." As Lynda Birke (2011, xix) argues, "It is extremely important that we recognize the involvement of nonhumans in the creation of cultures (human or otherwise), that we understand that they are not only 'good to think with,' but are also crucially partners *in* the making of our world." Methodologically and conceptually, there is significant power in recognizing that animals matter, warrant consideration, and have their own experiences, feelings, and needs. This expands on Josephine Donovan's (2007) call for striving to include the standpoint of animals in ethical deliberations.

The primary data collected is often from Canadian and US contexts but a number of cross-cultural examples are incorporated to capture and reflect a greater diversity of places and processes. Given the significance of animal work, I have sought to recognize a broad cross-section of workers, places, and processes. Each example considered warrants further, detailed consideration and I do not purport to have captured the full range of animal work present locally or globally. I also do not claim to speak for animals. Animals have their own ways of communicating; people need to try to understand them. I have sought to amplify, translate, understand, share, and propose, and to help create the conditions for more looking, listening, learning, and understanding.

By pursuing an expansive project and proposing a conceptual framework, this book offers different insights than those that can be induced from a locally rooted, detailed study. Such studies are important and my hope is that this book and the ideas herein will offer insights and intellectual tools of broad and specific applicability. Many of the ideas, organizations, and issues raised warrant deeper study and more illustrative examples. Overall, this book is not a sweeping generalization, an attempt to offer (allegedly) universal truths, or a detailed, micro-study. Rather, it draws from theory and empirically rooted evidence, and I share analysis, provocations, and a set of conceptual tools. In order to advance understanding and encourage new areas of inquiry, I have opted for an ambitious scope, and to introduce a broad cross-section of material. In many ways, this book is only a beginning, an attempt to assemble a cross-section of ideas and posit a series of intellectual concepts and political options.

Contextualizing Animal Work

Although thorough, historical examination of each of the local economies and communities considered is not possible in this text, the data are still understood in context. Put another way, the material is contextualized in time and space, and the most significant political, economic, cultural, social, and environmental dimensions are recognized when crucial. Context matters when trying to understand the social realm, including the many facets of animal work (Hedenborg 2007, 2009; Hedenborg and White 2013). For example, take a horse pulling a cart. This could be on a city street a century ago; in a rural or urban community today; near an Amish or Mennonite village; in a place of tourism like old Montréal, Niagara-on-the-Lake, or Central Park; inside a racetrack; or outside a royal wedding. The horse pulling the cart is only one piece of the story. To fully understand, you need to pay attention to the act and the actors, but also to put both into context.

Individuals' lives, workplaces, and local dynamics also need to be situated within their larger political and economic contexts. Although it is a recent socioeconomic invention, capitalism is the dominant, transnational global economic system today. It has a significant impact on human and nonhuman lives and deaths, and thus is crucial to understanding animal work. Capitalism is defined by two core features: production for profit and wage labor. Put simply, a small percentage of people own or run productive infrastructure, and most people work for those who do in exchange for a wage or salary. Countries have public sectors that exist simultaneously, are funded through public resources (taxes), and within which services and products are delivered for the public good. The relationships between the public and private sectors vary depending on the national or regional cultures, political parties in government, and the power of economic interests. Some societies have more active, public regulation and redistribution of wealth. Others follow a more dogmatic, savage, neoliberal agenda, which prioritizes deregulation, privatization, and the minimizing of public supports and social services for people and communities. In these latter cases, wealth and capital become polarized and the society more unequal. Animals, people, and all three types of animal work are shaped by these socioeconomic dynamics. Gender and class, as well as race, have been and continue to be particularly relevant to spaces of animal work (Gaynor 2007).

Moreover, capitalism is more than an economic system and mode of production. It is also a cultural and ideological project characterized by the pursuit of profit, competition, individualism, and the privileging of business ideas and priorities, often above other social, environmental, or ethical considerations. It is a powerful engine for economic growth and ideologically framed as a system of opportunity, merit, and choice, although, of course, people are very unevenly able to make choices about their work-lives therein. Like all economic systems, it is socially constructed, and the variations outlined shape and are shaped by cultural ideas. Processes of commodification are also central to capitalism, as products, people, and other living beings are transformed into commodities to be bought and sold.

When most people think of work, because of the normalization of the capitalist imperative, working for pay is top of mind, even though it is a recent phenomenon in the longer history of human societies. Waged/paid labor or income-generating work is very widespread today, and central to the study of animals and labor. Yet there are two other types of work that are also relevant. Unpaid work is, as suggested by its name, done without monetary remuneration. Unpaid work is a broad category. Most significantly, it includes domestic work in homes and all tasks and labor processes that are central to sustaining individual people and whole generations. Unpaid labor is also found in places of paid employment, through uncompensated overtime, extra tasks like planning and coordinating workplace social events, unpaid internships, or other forms of wage theft. Various forms of volunteer work, including community service and activism, are also examples of unpaid labor.

The third pertinent type of labor is subsistence work—that which is done to meet basic needs, stay alive, and literally subsist. As noted, subsistence work was the pillar of human societies for most of our history. It has been largely, although not entirely, replaced at the societal level, but it remains relevant to understanding particular individuals and groups, and to thinking more broadly about what every individual living being does for her/himself and, often, for others. Accordingly, there can be overlap among these three types of work, and examples of this connectivity are noted. Work is defined and understood in different ways within and across communities. It has multifaceted meanings and implications for individuals, groups, species, and the planet, and it can be rewarding, monotonous, troubling, insufferable, dangerous, lethal, transformative, and/or inspiring. But in every time and place, work is essential; so is animal work, and it is overdue for thorough analysis and greater care.

I

The Work Done With/For Animals: Daily Work and Labor Processes

Hundreds of millions of people around the world work directly with animals in many different ways. Across contexts, daily work tasks and whole occupations involve continuous engagement with animals. Such workplaces and relationships are fruitful places to understand the construction of human-animal relationships, conceptually and materially, and how interspecies dynamics are situated within larger socioeconomic, political, and cultural processes and systems. In that vein, Molly Mullin (1999, 219) argues that there is value in "continuing to consider humans' relationships with other species in relation to specific cultural and historical contexts and the ways in which such relationships are influenced by humans' relationships with other humans." The human work done with and/or for animals is the focus of this chapter.

Trying to identify the breadth of work done with animals is a major task, particularly if thinking cross-culturally and globally; thus descriptive categories are a useful starting point. Broadly speaking, work with/for animals can fit into the categories of service, law enforcement, military, health care, education, sport, entertainment, tourism, transportation, agriculture, food, resource extraction/mining, retail, training, research, welfare, and conservation. Undoubtedly there is overlap among these categories, as well as diversity in terms of the types of work being performed in each sector. For example, as part of its law enforcement work, a police canine unit does explosive sweeps, drug and/or weapon detection, tracking of people (suspects, seniors with dementia who have wandered, etc.), crowd control, education/training, and community relations. Health care work means the care of animals directly through veterinary work, and the therapeutic use of animals to provide care for people. Research work is also a broad umbrella under which studies can be conducted in naturalistic or laboratory settings, on captive or wild animals, and/or be noninvasive or about actively testing a product or procedure on an animal's body (known as vivisection). A primatologist or biologist with a doctorate degree observing a troop of wild apes in central Africa and a technician making slightly above minimum wage applying cosmetics ingredients to rabbits inside a laboratory in a North American city are both examples of research work. Similarly, agricultural work can include subsistence and/or poverty-income earning ploughing, growing, or husbandry; can be done on small, family-owned farms; or can mean waged work for large agricultural corporations. There are also dozens of separate occupations that fall under each sector. To take even a defined, confined space like a veterinary office, people work as veterinarians, veterinary technicians, veterinary nurses, lab technicians, animal care technicians, receptionists or clerical workers, and cleaning staff. This does not include peripheral or "spin-off" jobs in transportation, laboratories, and so forth. Thus, the descriptive categories are a place to start, but more specifics are needed in order to see a fuller picture.

Scholarly research helps deepen our understanding of the work done with/for animals. Without question, human labor done with animals is the subtype of animal work that has received the most scholarly attention. Sociologists, in particular, have examined human-animal work relationships, although the study of multispecies work still comprises a very small proportion of the total collection of research in the sociology of work. It is more common for a sociologist of human-animal relations to consider work than it is for a sociologist of work to incorporate animals into her or his research. At the same time, within the broader, interdisciplinary literature focused on animals, there is still only a small body of research that considers labor dimensions from any angle. In some cases, work may be present in the research site and constitute a small part of the data presented, but it is not prioritized and/or is subsumed behind other topical or conceptual emphases.

By employing ethnographic methodologies and drawing on symbolic interactionist approaches (put simply, that means an interest in how identities and relationships are constructed through interaction), a few researchers emphasize how human workers engage in meaning-making and continuously construct and/or contest boundaries between concepts such as human and animal, clean and dirty, and worthy and unworthy (Arluke and Sanders 1996; Birke, Arluke, and Michael 2007; Hamilton and Taylor 2013; Hamilton 2007, 2013; Sanders 1999). Workplaces examined include veterinary practices and clinics, laboratories, farms, and documentary filming sites. These studies offer locally rooted and detail-oriented data about the specific contexts being studied, and highlight the active, social production and reproduction of concepts and relationships, thus challenge the perception that conceptual categories are fixed and pre-prescribed.

Another small but important body of literature also draws on sociological traditions, along with neo-Marxist theory, to concentrate on labor processes. The core elements of Marx's theory of the labor process are purposeful activity, objects of work, and instruments of work, the latter being akin to a tool between the worker and the object of their labor. Labor process scholars tend to use this primarily as a springboard, and emphasize broader dimensions such as how work is structured, organized, managed, and, to some degree, gendered and racialized; how workers experience daily tasks and workplace hierarchies; and how individual and collective acts of resistance are pursued and to what ends. A handful of labor process scholars are interested in unpacking the work and work relationships involved when animals are present. They pursue what I would call an interspecies labor process approach, although it considers human interactions with animals in work relationships, but not animals' perspectives or work. Interestingly, most of this research centers on work at race tracks and/or in horse stables, but some also looks at farming. Ethnographic methods are often enlisted by these researchers, and/or interviews with workers are used as a data collection strategy. Scholars pursuing this kind of research have emphasized issues such as the changing and enduring nature of gender relations, industrial relations and worker voice, and how human-animal relationships influence the work and forms of agency (Butler and Charles 2012; Butler 2013;

Hedenborg 2007, 2009; Larsen 2006a, 2006b; Miller 2013a, 2013c; Wilkie 2010).

Both of these approaches to the nature-labor nexus provide helpful data and insight, and I suggest that there is value in combining their emphases in order to interconnect symbolic dimensions like meaning-making with more decidedly labor-focused inquiry. This suggestion builds on the arguments of anthropologist William Roseberry (1989) who sought to transcend yet another dichotomy, one between material/political economic and cultural/symbolic analyses. Although the two approaches to work in multispecies species outlined draw on different intellectual genealogies, they are not incompatible. Moreover, symbolic interactionist researchers do not completely avoid discussion of political economic matters, and labor process researchers do not entirely eschew consideration of meaning-making. In addition to a shared interest in work, both also begin with an interest in empirical evidence, qualitative data, and specifics. Rhoda M. Wilkie's (2010) ethnographic study of the work done with animals intended to become food provides an example of how more material and political economic questions can be considered along with experiential and symbolic understandings and dynamics, and her work will be revisited throughout this book. By recognizing both broad spheres, we gain a more holistic understanding and are able to contextualize symbolic dimensions within their political economic locales. We also can see how the tangible dimensions of work shape, and are shaped by, the realm of ideas. Knowing more about the political economic dynamics and the structure of work allows us to better contextualize the perspectives of workers, and more fully understand how and why particular ideas and meanings are constituted and/or contested.

Interestingly, despite shared or similar structural dynamics and labor processes, specific workplaces and workers can still differ in noteworthy ways. Arnold Arluke and Clinton R. Sanders' (1996) examination of two university-based, primate research laboratories in the same city, reveals clear differences between the two sites. These include the motivations of the workers present, their attitudes toward the work and the animals, and the hiring processes. In one lab, most workers demonstrate a more utilitarian approach to the work and do not express any connection with or sympathy for the animals. In the other, there is more of a culture or ethic of care, as relationships with individual animals are built and nurtured, the animals are named, and workers uncomfortable with testing seek to make daily life as pleasant as possible for the primates. Gendered dynamics are at play, and Arluke and Sanders (1996) refer to the workers in the rougher lab as "cowboys." The individuals responsible for hiring and their ideas about the work, workplace, and animals have also shaped who else is present and involved. The supervisors have exercised differing degrees of agency to create the conditions at work, particularly how the processes and participants are understood, and how the lab operates, thereby shaping people's experiences of work and the monkeys' daily lives, to some degree (although in both cases, the testing continues). Thus, even in a single city, the evidence makes clear that generalizations cannot be made about many aspects of work in primate labs. This reaffirms, among other things, the value of local and specific data, comparative analysis, and consideration of both structure and agency. Work and workers are structured, organized, and constrained, but workers are active social agents who can cooperate, question, contest, resist, and change workplace ideas and relations.

Working with Animals for a Living?

Undoubtedly, questions of pay and income are crucial to understanding all animal work. People's work with/for animals is very differently remunerated, but much of it is poorly paid and is quite precarious, which means it is also contingent, erratic, insecure, and often part-time. Precarious work is both economically and socially devalued. For women and racialized workers in particular, such conditions are not new, but precarious work is increasingly common and widespread in countries like Canada, the United States, and Britain, among others (Kalleberg 2011; Vosko 2000, 2006; Vosko, MacDonald, and Campbell 2009). It is also helpful to understand how jobs fit within their larger social and political context. This includes determining whether other provisions for workers' well-being are available or guaranteed, such as health benefits or insurance, paid sick days, overtime pay, and other legal rights and protections. In many places across Canada and the United States,

for example, agricultural workers are legally prevented from joining unions, and some are even excluded from workers' compen-
sation provisions or other labor laws, regardless of whether they
are citizens, permanent residents, or migrant workers. Many jobs
with riding or racing horses, even if the racetrack is in an urban
setting, is classified as "agricultural" work under the law in a number of jurisdictions (Cassidy 2007). These dynamics have tangible
impacts on people's lives. Eddie Sweat, groom to the decorated and
accomplished racehorse Secretariat, died in poverty, a reflection of
a few factors, including an industry in which most people toil for
very little pay (Scanlan 2006).

Jobs with animals that are materially more comfortable and
secure are primarily in the public sector and are often unionized.
This includes subsectors like policing and natural resource management, although there are a range of pay scales within this kind
of work. Veterinarians, whether working in private practice or the
public sector, are among the best paid animal workers, while veterinary technicians, who are required to successfully complete training and education programs, still earn only modestly more than the
minimum wage in many countries. The incomes of small business
owners or employees in areas like grooming, kennelling or day care,
training, and dog-walking can be volatile and range a great deal,
and, although the "animal business" sector is continuously growing,
small businesses of all kinds, especially newer ones, have very uneven
survival and success rates (Fisher and Reuber 2010). Horse riders
in sporting industries like racing, show jumping, polo, and rodeo
also have very divergent incomes and political economic situations
(Coulter 2013c). In these cases the most successful examples may
become household names, but most people will earn either a low or
modest income. As farming is restructured and family-owned farms
are replaced by large, agribusinesses, the already modest incomes of
farming people become increasingly precarious. Most farm families
in countries like Canada have at least one person working for wages
off the farm, and as agricultural corporations buy up farm land and
replace a higher number of smaller farms with massive operations,
the jobs offered are usually insecure and poorly paid.

Some may suggest that the chance to work with animals is payment enough. Alternatively, some particularly militant labor advocates will level fierce critique at even nonprofit organizations that

enlist volunteers. In certain situations, donating one's time to work with/for animals is laudable, as well as nonmonetarily rewarding. Moreover, particularly not-for-profit organizations strained for operational resources may benefit greatly from voluntary labor and would be unable to deliver basic levels of service or care without unpaid workers (unless they were provided with new sources of revenue). There are also examples of well-paid animal workers, such as veterinarians, who donate their labor to organizations providing care in poor communities, locally or internationally, or discount their services for people of lower incomes or who are doing rescue work, thereby taking a voluntary pay cut of sorts. This kind of "pro bono" work is commendable.

As is the case across sectors, voluntary labor with animals thus needs to be analyzed in context, rather than uncritically glorified or condemned. For example, not everyone is equally able to volunteer their time. People with higher class positions and reliable family incomes are more able to genuinely volunteer. Given the gendered and racialized makeup of class hierarchies, there are clear inequities in terms of who is able to freely donate their time and effort without worry, and thus be seen as generous and community-minded. More to the point, most people simply cannot work for free. The median income in a wealthy country like Canada, for example, is around $30,000 per year for individuals and $70,000 for families (Statistics Canada 2011b). This means that half of working-age Canadian people and families earn less than these amounts.

At the same time, volunteers may lose interest, be unable to continue, or only have a few hours to dedicate. In workplaces with a mix of paid and unpaid workers, there are also implications for waged employees' workloads, as they are responsible for coordinating, supervising, and monitoring volunteers' work. Particularly as many governments cut back on public services, political leaders may promote or mandate volunteering as a replacement for work that was previously done for pay. In other words, what used to be someone's job becomes a "volunteer opportunity." At the same time, some jobs are cut and then not even formal volunteer positions are put into place. Undoubtedly this is problematic, and research has consistently found that it is especially women who take up the slack to try and keep valuable programs and services going, thereby adding to their already large unpaid workloads. There are also generational dynamics in play. In some places, high school students are required to work for free (i.e., "volunteer") for a set number of hours in order to graduate. After graduation from high school or post-secondary education, some young workers are pressured to seek out unpaid internships in order to try and gain contacts and get a foot in the door toward paid work, regardless of whether this is ultimately successful or not. This pattern is not unique to young workers either, and older people who are replaced or terminated may feel and face similar pressures as they attempt to gain employment in a new or different line of work.

This larger political context complicates ideas of unpaid work and volunteerism, the latter intended to be about community, altruism, and service, but currently something much more complex. This discussion is not intended to suggest that volunteer work is not helpful, laudable, or even essential, as some animal work spaces rely exclusively on unpaid workers and some animals are only alive because individuals have donated their time, often without fanfare, to provide care. Rather, the role of unpaid labor needs to be understood in context, and should not be seen as a replacement for paid work with animals. People need incomes to survive and maintain a decent quality of life.

Indeed, millions of people still make a living—or struggle to do so—by working with/for animals. People's perceptions of animals, jobs, and interspecies relationships are shaped by pay and experiences at work. Moreover, political economic factors influence how animals are treated. Clay McShane and Joel A. Tarr's (2007, 51) analysis of the teamsters who drove teams of cart-pulling horses in US cities during the rise of the industrial era reveals clear differences in how the men treated horses, but the authors largely agree with the Massachusetts Society for the Prevention of Cruelty to Animals' summary: "If the men are on good terms with the employer, the horses are usually well treated, whereas if the men are dissatisfied, the horses are always badly treated." This is a finding of enduring and broader significance, one that I will revisit in later chapters.

At the same time, pay and material conditions of work are not the only determinants of job satisfaction, nor is there a deterministic, causal relationship between pay and what people think. But the conditions of work are important and shape how people feel about themselves and their labor. This, in turn, affects not only people's well-being, but also the lives of the animals under their care. Similarly, working conditions influence workers' tenure and rates of turnover, which also affect animals' lives. This is both because of the relationships animals can build with familiar people, and because turnover affects workplace operations and interspecies service delivery. In racing and show jumping stables alike, for example, high rates of turnover (which can top 100 percent annually) are common, and undoubtedly the low-wages, in combination with the long hours and demands of the work, contribute. A high rate of turnover means different workers keep coming and going; new employees need training, lack the knowledge of experience, and are unfamiliar with the workplace particulars, especially the animals. When working with individual animals who have their own minds, personalities, and preferences, such personal, experiential knowledge often proves invaluable.

Low pay is often interwoven with the feminization of work, and this is true for a number of jobs with/for animals. Work, overall, can be gendered in two ways. The first is because of who does the work most often. If an occupation is numerically dominated by women, it is "feminized"; if mostly men do the work, the job is considered male-dominated. Second, if the characteristics and expectations of the work are associated more with one gender, the occupation is gendered masculine or feminized. In other words, even if men numerically dominate a position, the work can still be considered feminized if it involves caring, emotionality, service, and/or other attributes that have been socially ascribed to women's roles and characteristics. Accordingly, scholars like Donna Haraway (1991) and Janine Brodie (1995), among others, have argued that the growing pool of precarious, insecure, and marginal jobs (historically associated with and done by women and workers of color), which are replacing stable, full-time work, should be understood as feminized. Work that is feminized is often dismissed as allegedly low skill or unskilled (see, e.g., Coulter 2014b). However, feminist and labor researchers have contested this perception, illuminated the range of skills involved, and challenged how labor processes are understood. This includes thinking politically about individual and collective bodies at work, an angle that undoubtedly applies to animal work contexts.

(Human) Bodies of Animal Work

Janet Miller's (2013a) unpacking of the work done by stable staff with/for race horses offers valuable insights of specific and broad applicability for building a nuanced and multifaceted understanding of work with/for animals. Miller observed and interviewed 90 stable workers/grooms, people whose daily work revolves around race horses in Britain. These are low-paid positions across countries, and of the workers interviewed by Miller, about 57 percent were men and 43 percent were women. Although a majority of the workers are men, for the reasons outlined above, particularly the low pay and caring requirements, this work could aptly be considered feminized. In countries like the United States, such work is also often racialized, and US citizens of Caribbean and Latin American descent, as well as migrant or undocumented workers from both regions are well-represented among the ranks of grooms and stable staff in show jumping as well as racing stables (Castañeda, Kline, and Dickey 2010).

These people work daily in stables doing a range of tasks including feeding and watering the horses, cleaning the stables and stalls, and tending to the horses directly. Such work involves continuous interactive work including grooming, tacking up and removing tack (equipment like saddles, bridles, and boots), and basic medical assessment and treatments. Some also work as exercise riders but many do not. The work is difficult, poorly paid, and workers may not be aware of their basic legal rights (Brooke-Holmes and Calamatta 2014). Lawrence Scanlan (2006, 55), in his nuanced portrait of Eddie Sweat, the groom for Secretariat, synthesizes it well: "[W]hen the photograph has been taken, the owner goes back to his or her fancy box, the trainer and jock[ey] move on to the next race, and the lowly groom, horse in tow, does his or her duty. Walk, bathe, feed and water the horse, muck out his stall, pick his feet, clean his tack, blanket him, rub him down, load him on the van. No one understands that horse better than an astute and caring groom, and no one gets less credit."

Indeed, a basic description of the daily tasks does not fully capture the complexities of the labor processes involved. Recent research has emphasized the importance of seeing how bodies are involved in/at work in a range of ways. This is a different emphasis from research that focuses primarily on how individual bodies are experienced, and is instead more of a sociological and labor-focused lens. Carol Wolkowitz (2006), in particular, developed the concept of body work to refer especially to the work done to the bodies of others, but it can also refer to work people do to/on themselves. This can include intimate and sexualized work, as well as various health, sporting, and beauty occupations. It can also refer to specific workplace tasks even if an occupation is not focused on bodies or on continuous body work. In other words, we can identify specific body work undertaken on occasion, as well as entire occupations organized around the performance of body work. As Rachel Lara Cohen, Kate Hardy, Teela Sanders, and Carol Wolkowitz (2013, 4) note, all labor involves bodies, but by thinking about the concept of body work, we are prompted to see bodies as "materials of production." Because of its breadth, body work can mean very different things and can be used by researchers in quite distinct ways. External expressions of body work, that is, doing things to the bodies of others are the most explicit examples and easiest to identify. When workers are working on their own bodies, the concept becomes more challenging and the boundaries less clear. In this chapter, the use of body work is more straightforward, but in subsequent chapters, its potential uses as well as the challenges of enlisting the concept are discussed in more detail.

Employing a gendered labor process approach, Miller (2013a) enlists the idea of body work as a central frame for understanding the work of stable staff and grooms in the "production" (and reproduction) of race horses. Horses are born with the ability to run, but through multifaceted physical and intellectual training (or socialization), horses become formal racehorses. The (re)production of racehorses also includes the political economic dimensions of actively engaging in the breeding and sale of horses. In horse stables, human workers are responsible for working on/with two bodies—human and equine. The body work people perform on themselves makes it possible for them to work on/with horses. The humans, especially those given riding work, are required to be light but fit and strong (Miller 2013a). They need to be able to carry out physically demanding tasks including moving hay bales and water buckets, and cleaning stalls, but weigh the same or little more than jockeys (who generally must weigh less than 120 lbs/55 kgs). Proper grooming of horses involves a measure of body strength and flexibility, as well. At the same time, interactive work with horses is not always or even often about strength, and it regularly involves precision and delicacy. In the case of stable staff or grooms, the body work they perform daily is physically demanding in both its rigor and intricacies, and made even more so due to long hours and early mornings. Stable staff must do work on their own bodies to maintain the proper level of fitness and strength, while also working on the bodies of horses to ensure they are clean, healthy, and fit.

It is thus important to recognize that interspecies body work involves a complex and negotiated physical, interactive, proactive, and reactive process. In most horse work contexts, although humans have domesticated equines, keep horses under conditions and for purposes largely of human's choosing, and may enlist tools of pain or punishment to instill their will, the daily dynamics at work are more complex than simple human over horse mastery. Horses are not usually physically overpowered. Instead, humans use their bodies to persuade, direct, assess, react, prevent, and understand. People do so with varying degrees of kindness and force. But among those engaged as stable staff or grooms, the low-paid workers tasked with the daily care and management of horses, force is not particularly effective or widespread. Many care for horses, literally and conceptually. Thus here, Miller's (2013a) analysis is again helpful as she enlists other linked but distinct concepts to help understand the complementary labor processes involved.

Work with animals involves presence; living beings of different species are physically present and involved. Animals can both feel and inflict pain (accidentally or deliberately), as well as pleasure. They are living beings with bodies, bodily needs and functions, minds, moods, and personalities. Interspecies work is a visceral, embodied experience, and when dealing with animals directly, physical interactions mean bodies are involved in the fullest sense. As a result, dirty work is also involved. Dirty work refers to work that is deemed degrading and/or undesirable. It often refers to work that is physically unpleasant, involves dealing with bodily fluids, excrement, and the like. It thus can refer to an entire occupation that revolves around regular completion of dirty work. Alternatively, there may be specific tasks that involve dirty work, even if the occupation itself is not considered "dirty work." For example, surgery may mean a doctor gets splattered with blood and is wrist-deep in pus, thus engages in some dirty work, but being a surgeon is not deemed undesirable or a dirty job.

Dirty work is clearly involved in many animal work contexts, including veterinary, grooming, and agricultural work. In stables, cleaning manure and urine-soaked straw or shavings out of horses' stalls, getting dusty and dirty from grooming, cleaning male horses' sheaths, and packing, pressing, or dressing wounds are commonplace practices. These kinds of tasks are widely seen by outsiders as examples of dirty work. However, not all stable workers, whether in racing, show jumping, or other equine industries, see these jobs as unpleasant or deem dirt to be a problem (Cassidy 2007; Miller 2013a). Some find great reward in seeing the fruits of their labor and in ensuring the horses' needs are being met as best as possible. Accordingly, this dynamic segues into another important concept: care work.

Care work is an established area of study focused on the processes and people who provide care, and it refers especially to labor in areas such as nursing and social services. The provisioning of care can be physical and/or emotional. Many whole occupations revolve around care work, but certain kinds of care work can be done as one component along with a number of other types of work. For example, a teacher's primary responsibility is education and learning—and she/he engages in education work—but care work is often needed and involved, particularly to help, comfort, and support troubled students, those facing multiple barriers, and those dealing with crises. Care work is provided through publicly funded organizations, or available for purchase through the private sector, as well as done without pay in homes and communities everywhere in the world. In all cases it is highly feminized, often racialized, and almost always devalued (Armstrong and Armstrong 2005; Armstrong, Armstrong, and Dixon 2008; Boris and Parreñas 2010; England, Budig, and Folbre 2002; Glenn 2010).

The concept of care work clearly applies to animal work contexts. Entire occupations in veterinary medicine revolve around care work, of course. Daily care work is also essential in any space where animals are kept to keep them alive and ensure that they fulfill their roles (whatever those might be). This can be a very instrumental or functionalist kind of care work premised on keeping the animals alive or in "good enough" shape. Alternatively, daily care work can be interwoven with a broader approach to caring, which takes into consideration animals' individuality, sentience, emotions, desires, and needs. Both individual and contextual factors shape how care work is conceptualized and performed.

Miller (2013a) notes that caring for horses is integral to stable labor, for example. The larger structure of racing makes certain interactive patterns possible while preventing others, but barn cultures and individual workers also shape local and interpersonal particulars. Basic essentials include ensuring that horses have adequate food and water, clean conditions, and healthy bodies in order to be able to race. The promotion of good health requires knowledge of what is normal for the animals overall and for individual horses. Potential issues must not only be recognized and identified, but also addressed, and future problems are also minimized through regular assessments and preventative measures. Therefore, an understanding and awareness of horses' bodies, as well as their moods, is essential. This requires specialized knowledge of equine anatomy, physiology, and behavior, as well as an ability to understand what horses may be feeling or trying to share. Consequently, relationships and experience are also key. Indeed, these dimensions are integral to communication work, another crucial, intersecting dynamic of the labor process in horse stables and across workplaces involving animals.

Communication work takes on a special meaning in interspecies relationships because humans and animals do not share a full verbal language, although some animals learn a small vocabulary of human words. As a result, effective understanding and exchange is not automatic, but rather requires continuous reflection, control, augmentation, and adjustments to promote understanding. Entire firms exist that specialize in communication skills and strategies to help people communicate with and persuade others more effectively. Communicating with other species is a challenge, but there are no high-paid consultants upon which to draw. Some animal behaviorists and/or trainers hone their abilities to understand and teach interspecies communication, particularly by studying the ways horses and dogs communicate with each other. Some self-defined animal communicators and psychics also exist. Moreover, the growing bodies of animal behaviour and cognitive ethology research are instructive. But most individual human workers develop their abilities to understand and share across species through experience. This involves recognition of the need to communicate and a willingness to try and build bridges. Moreover, human workers must not only develop their abilities to understand what animals are deliberately and unintentionally communicating, but also to respond. In this process, the content and tone of speech, eye contact, touch, visual and physical assessments, gestures, posture, body position, and other senses can all become involved (what some social theorists call affect). At the same time, people may communicate with animals unintentionally, as elevated heart rates, anxiousness, and other such physical signs can be understood by animals and betray our true feelings, which can in turn shape how animals feel.

In a horse stable, workers who have developed communication skills can read how horses are feeling based on how they behave and look. A different look in a horse's eyes, ear positioning, tail swishing, and other indicators such as an unwillingness to interact or a more subtle conveying of displeasure, can all be signs of physical and/or mental discomfort. Not only do stable workers need to understand the signs in a general sense, but they also must know individual horses, and actively take note of even small changes. The most cherished stable workers are those with this kind of horse sense, along with a personal commitment to the work and to the horses. A distracted, inattentive, or indifferent worker may miss signs. Those working with animals do so for different reasons, and have divergent feelings about what they do, and the limitations and possibilities of their daily labor. Undoubtedly, emotions are a complex and deeply significant aspect of animal work.

The Emotions of Work with/for Animals

The role of emotions at work is a well-studied area, particularly in the literature on service work. Emotions and work can be conceptualized as an umbrella under which distinct but related concepts fit, each of which can be relevant to different spaces or aspects of animal work. Emotions also intersect with how people feel about the animals with whom they work, and affect political action and agency. Core terms and processes will be introduced here, and revisited throughout the book because of their significance.

People's feelings about animals can influence their decision to pursue interspecies work and occupations. In fact, workers doing different types of animal work say that their "love" for animals inspired their employment choices, illustrating the concept of emotional motivation (see, e.g., Cassidy 2007; Sanders 2010; Taylor 2013). Stuart J. Bunderson and Jeffery A. Thompson (2009) discovered a recurring pattern among zookeepers, for example, who felt that their work was, in fact, their "calling." Particularly for interactive workplaces, a desire to work with animals on a daily basis motivates people to pursue specialized training, or to begin learning about the animals and industry while they are young. Similarly, young people's demonstrated concern for animals, evident in their compassion for stray or wild animals, and/or in the care they provide for animals can trigger a desire to pursue a career with animals and/or for adults around them to suggest the same. Even young people in poor, inner-city neighborhoods may envision a life working with animals because of films or books, interaction with animals, experience at a shelter, connection to someone who works at a horse racing track, or if they learn of one of the urban stables that continue to exist in places like Los Angeles and Philadelphia (Camarillo 2006). In such contexts, young, often racialized men speak openly and proudly about the sense of community and respect they gain from working with horses and horse people, crediting these interactions with keeping them alive, out of jail, or motivated to work with horses all their lives. Clinton R. Sanders (2010) found that an affinity for animals provides the impetus, but that a connection to someone in the field is important for gaining a genuine employment opportunity.

Emotional motivation and an interest in interspecies work can also shape specific career trajectories within larger occupational groups. In police forces, mounted and canine units are widely seen as coveted postings, accessible only to a minority of officers who meet the qualifications and have sufficient years of training and experience. This is despite the fact that most forces require horse stall cleaning and round-the-clock care for dogs, which always means at least some dirty work. Moreover, even among low-paid horse workers, I consistently heard that a desire to work with horses was the primary motivation for their paths. When asked about the hardest part of his job, one groom captured the views of many of his coworkers when he said: "The people. Everything about the horses I like. It's the people who cause me problems." Particularly among those able to choose their jobs (i.e., citizens not facing substantial barriers or discrimination), despite the low pay, long hours, and dirty work, some workers stayed in the industry for years, even if they moved between a series of different stables. Of course, others ultimately sought other work, highlighting a desire for a "non-horse job" and "things like weekends" or benefits, but many were adamant that they needed "trees outside my window and to see horses every day." In an interview, Brian Tropea, a lifelong horseman and general manager of the Ontario Harness Horse Association put it this way:

> You know, the old saying, "the outside of a horse is good for the inside of a man;" and I don't think that there could be anything more true than that. I remember when I was a kid, I'd have a bad day at school, somebody would be picking on me or something, and I wouldn't tell anybody in the house, but I'd go out to the barn and tell my pony, and I knew the pony wasn't going to tell anybody. And I really do believe that they are therapeutic animals, and they are underused [in therapeutic contexts] I think. There's a lot of people that aren't employable doing something else where they are in a cubicle and they got to deal with stress and everything else. But they work fabulously working with animals, you know? Never had to deal with bureaucracy or anything, they just had to make sure they were there when the horse was hungry, and sick and tired, and that's their life. That's what they know and that's what they love. And like I say, a lot of people just aren't hardwired for that 9–5 and they are hard-wired from six in the morning, you know? They get up and they do it and they complain about it sometimes that they're not making enough, that they can't afford to do anything, but they're productive citizens…I mean, I know a lot of people in the industry that are second, third, fourth generation horse people that could have went on to school, but [they] knew that this was what they wanted to do.

At the same time, some stable staff, especially younger women, leave groom/stable work to train for positions like veterinary technician. Although such positions usually do not pay much better, the more reliable hours and professionalized contexts are appealing, and still allow for direct, ongoing interspecies interaction, albeit with different animals on a daily basis. Veterinary technician and nursing work has long been feminized numerically, but across a number of countries in the global north, veterinary doctor positions are also becoming increasingly staffed by women and a large majority of current veterinary students are women. Tuition fees are high for these programs, so while gender is less of a barrier, class and income continue to affect whether those motivated toward animal health care can become doctors of veterinary medicine, or whether programs like veterinary technicians with lower fees are more within their financial reach. Moreover, gender politics are not as simple as numbers, and how or whether this gendered shift changes the profession is still being determined. Early research suggests that the influx of women has not changed the gendered expectations of the jobs significantly (Irvine and Vermilya 2010). As Susanna Hedenborg argues (2007, 2009), it is important to recognize continuity and change, and how multispecies specifics are affected by—and can affect—the larger sociopolitical, cultural, and economic context, including gender norms.

Once people are involved in working with/for animals, emotions continue to play a central role in their work-lives across contexts. The concepts of emotional labor and emotion work can come into play in interspecies workplaces, as well. Emotional labor is an idea developed by Arlie Hochschild (1979, 1983). After studying flight attendants' work, she argued that these women controlled and performed emotions as a regular part of their jobs in order to deliver the best service, and thus make the most profit for their employers. Emotional labor refers especially to the visible, performed, outward aspects of the process, and often in for-profit contexts, although it has been and can be applied to more than waged employees. The self-employed or small business owners may perform emotional labor regularly, for example.

At the same time, people often engage in additional, internal work to control their own feelings. This is often known as emotion work. Hochschild originally argued that emotion work was done specifically in noncommercial/private spaces like homes, but recent writing has recognized that emotion work, and especially emotional management, is done in formal workplaces as well. Veterinary practices are key sites where both emotional labor and emotion work are continuously required, because seeing animals hurt, sick, and being euthanized is a recurring part of the job. Veterinary educational programs may formally or informally prepare students for these and other emotional demands of work in the field (Hazel, Signal, and Taylor 2011; Vermilya 2012). In his ethnographic study of a veterinary office, Clinton R. Sanders (2010, 248–9) highlights the emotional dimensions of the work as the most challenging for the veterinary technicians who were the focus of his study, and for himself as a researcher. He puts it frankly, "What I found most difficult to deal with—and never became entirely insulated from—was the sickness and death of the animals brought to the clinic and the intense emotional pain experienced by their caretakers." In other words, personal pain results from both seeing animals in distress or at the end of their lives first-hand, and from seeing the effects this has on the people who love the animals. Veterinary workers are responsible for maintaining professionalism, while expressing concern for the people whose animals are suffering or being euthanized, regardless of their own emotions and feelings. They are to engage in the emotion work necessary for the successful performing of emotional labor on a regular and even daily basis, and workers develop and hone different strategies depending on their circumstances and needs. As a result, the concept of emotional management is also interconnected with emotion work in interspecies workplaces.

Depending on their context, jobs, and degrees of emotionality, much like nurses in palliative care and comparative fields, people working with/for animals employ a range of emotional management strategies to deal with pain and death. Some seek to distance themselves and detach, something suggested to Sanders (2010) and to new veterinary technicians by their coworkers, along with the importance of prioritizing the pet owners' feelings. Such emotion work is seen as central to creating an environment that is as supportive as possible, not one that magnifies sorrow through the visible sadness of staff. A focus on precision and accuracy is another strategy used by workers (Birke, Arluke, and Michael 2007; DeMello 2010). People work to ensure that the fatal needle is administered as correctly as possible to minimize the animals' pain and stress, for example, and emphasize the need to fully follow proper procedure in all instances. The overall alleviation of pain is another dimension people self-emphasize, purposefully focusing on the fact that the animal was suffering. Some workers also seek to bolster their detachment at work with strategies of escapism or coping at home. Of course, certain people simply cannot develop effective coping skills and their mental health may suffer, and/or they will leave the profession (Birke, Arluke, and Michael 2007; Sanders 2010). A growing collection of research has found that levels of depression, anxiety, and suicide are notably higher among veterinarians than the general population across countries, as well as greater than those of other health care workers (Bartram and Baldwin 2010; Bartram, Yadegarfar, and Baldwin 2009; Skipper and Williams 2012). Researchers hypothesize that a few causes are influential, but the high frequency of euthanasia is consistently identified as a contributing factor.

Not everyone who works with/for animals is an animal lover, however. Among certain workers, emotions are not a major motivator for their work, nor do they even necessarily have positive feelings about animals. Some do not have much choice about where they work and what they do, particularly if they live in an economically depressed community where there are few options. At the same time, for some, work with animals is just a job and a pay check. For certain people, the interspecies dimensions are, in fact, a negative part of their work. For example, among the primate lab workers (and men in particular) interviewed by Arluke and Sanders (1996, 109–10), one compared his job with animals to food service: "The first month you work here it's fun. It's like working in a pizza place. The first day it's great, you eat pizza all day long. The second day, you're kind of sick of pizza. Just like here, I work with the animals all day long. Everyone has free time, but playing with the animals isn't all that great. I would rather go out with some friends and party than stay and play with an animal." Arluke and Sanders (1996, 110) found that for certain workers (namely the "cowboys"), "not only did relationships among people take precedence over spending extra time with animals, but the commitment of cowboys to the animals was weak enough on occasion to compromise basic veterinary care. Doing 'just a job,' they sometimes failed to complete their duties because they were 'in a hurry to get out of here.' One caretaker, for instance, went home an hour early, leaving ten animals with empty water bottles." These examples are important to note, and workers' feelings vary depending on a range of factors, including the type of work they do. That said, the research makes clear that many and potentially most people who take up interspecies work do so because they have positive feelings toward animals.

Without question, emotions and emotional work are central to much interspecies labor and manifest in different ways. In many interspecies workplaces, the joy of working with animals inspired people's workplace direction, and such feelings outweigh the sorrow of animals ultimately being sold or killed. The connections formed with animals, and the ability to help them are recurring themes shared by people in a diverse, cross-section of animal work spaces; the specific roles emotions play are shaped by the context and the social actors involved. Certain occupations are undoubtedly more rewarding, while others are more emotionally complex, such as those involving animal pain and death, as well as cruelty to animals. Animal cruelty investigation work is highly challenging. Workers must be part police officer, part nurse, and part social worker, and manage intense emotions. Arnold Arluke (2004) argues that workers in these positions tend to develop "humane realism" to cope with witnessing the horrors of the abuse and neglect of animals, a widespread lack of recognition for their work, the frustrations of understaffing, and what can be a lack of motivation on the part of Crown prosecutors to pursue cases involving animal harm (when laws do exist). Workers who do investigative work can become frustrated by a range of social, logistical, and organizational factors that limit their ability to prevent cruelty in the first place. Thus Arluke (2004) notes that while the development of humane realism helps workers cope and recognize what they can and do accomplish, they are limited in their ability to challenge the conditions that cause their frustrations. This is one of many challenges that cloud the literal and intellectual terrain of animal work, even in professions that involve helping animals. I will further revisit these issues and more thoroughly discuss labor that revolves around political work for/with animals in the third chapter. Now I turn to an area of the nature-labor nexus fraught with tensions and deep emotional complexities: agriculture and food.

Work-Lives and Deaths

Because many humans choose to consume meat and other ani-mal-derived products, a great deal of work is done with animals intended to become food while they are alive, and as and after they are killed. Within this broad category, there is noteworthy diver-sity in the type and structure of labor performed, and among the people who engage in it. Globally, 1.3 billion people are involved in agriculture, most of whom are poor, rural people working with only a few animals (Steinfeld et al. 2006). The Food and Agriculture Organization (FAO) (2009) estimates that close to one billion, or 70 percent of the world's most impoverished people, rely on livestock for their livelihoods. About two-thirds of poor livestock keepers are women (Ibid.; The Brooke 2014). In these contexts, work is often about subsistence, and poor people often raise animals for their milk, eggs, and/or meat, and/or use animals to help take products to a local or regional market. It is physically difficult and volatile work, but involves very small-scale activities. This was the norm in most parts of the world for many centuries before agriculture became increasingly industrialized (Schwartzman 2013). But even the early roots of ranching in countries like the United States and Canada were infused with differential power relations and both symbolic and material significance (Nibert 2013). Jean O'Malley Halley (2012, 15) argues that cows "hold a central, symbolic place in a national [US] story of origins... The story of the meat industry mirrors the story of national economic development. It is a story of white colonization of what became the United States, the interac-tions between small farmers and developing industrial meat busi-nesses, governmental intervention in and support of businesses small and large, technological development and industrialization and the national consumption of mass-produced food."

Today, in countries like Canada and across much of the global north, the number of people who work in the broad sector of agri-culture is declining, and it has been for the last few decades. This is because of changes to both how animals intended to be food (as well as leather, fur, and other products) are kept when they are alive, and how they are killed and then processed (Shiva 2000; Stull and Broadway 2013; Torres 2007; Twine 2013; Weis 2007,

2013). The image of a small or modest-sized family-owned farm that dominates public imagination and popular culture constructions is increasingly inaccurate (Pini and Leach 2011). These kinds of farms are being replaced by larger "factory farms" or agribusinesses, where crops are grown or very high numbers of animals are kept. These facilities may also be called intensive (or industrial) livestock operations or concentrated animal feeding operations. As Josh Balk (2014, n.p.) puts it, "Farmers' pride in animal husbandry has [largely] been replaced with agricultural systems treating animals as if they were machines in an assembly line."

In 2011, there were only about 205,000 farms in a country as geographically large as Canada. This is a 10 percent decrease from five years earlier, and a substantial reduction from the more than 700,000 farms in operation in 1931, and the 500,000 in 1961 (Statistics Canada 2011). Since 1991, the number of farm operators has decreased by about 100,000 people to just over 293,000 (Beaulieu 2014). About 85,000 farms grow crops, 37,000 are for cattle who will become meat, 12,000 are for cows who will produce milk (and then become meat after a few years), and 8,000 farm fruit. The rest are for chickens, turkeys, pigs, other livestock, vegetables, or some combination, such as a mixed animal-crop farm. The large majority of animals producing or intended to be food live indoors in massive, uniform, often windowless industrial shed-like facilities, usually built in rows. Some can be found down very quiet gravel roads, others are kept behind tree lines, and certain are in plain view. Many people think of red barns painted with farmers' last names or pastures as where animals live, and while these do exist, they are less and less common. Someone driving down an average country road would likely be able to see cattle grazing in herds (often brown or black in colour) who are intended to become beef, the occasional chicken around someone's house, and perhaps some grazing sheep, goats, or cows (usually with black and white markings) used for milk production, although many of such cows are housed indoors year round. The majority of other species, including at least 90 percent of chickens, turkeys, pigs, geese, rabbits, and so forth, are always kept inside. Fur farms also house all animals inside. In the United States, Canada, and a number of other countries, far more chickens are killed than any other species.

It is as food for humans that most animals live and die, and despite the smaller number of farms and farmers, more animals are being raised, killed, and consumed. The Canadian Federation of Humane Societies (n.d.) calculates that close to 700 million farm animals are killed every year in Canada (despite there being less than 100,000 farms raising animals for food). The number of animals involved with food production is dramatically higher than the 14 million companion animals who share people's homes and lives, and the approximately 3 million animals (especially mice, rabbits, and dogs) used in Canadian laboratory research. The total number of wild animals is very hard to determine. The Unites States Department of Agriculture reports that about 9 billion animals are killed annually within US borders, but that figure excludes horses, rabbits, fish, and crustaceans, so the total is much higher (Humane Society of the United States 2014). The numbers climb by tens of billions when considering the global situation.

The facilities where animals are sent to be killed have also changed in ways that impact people and labor processes, as well (Lee 2008). Amy Fitzgerald (2010) has traced the historical progression of slaughterhouse organization, which originally involved private killing of animals. Slowly these processes became centralized and regulated, and the slaughterhouse became a specific institution in the early nineteenth century. Further centralization and industrialization progressed from there, with the massive Union Stock Yard complex opening in Chicago in 1865, surrounded by slums where 60,000 workers and their families lived (Fitzgerald 2010). In fact, Henry Ford says he took inspiration for assembly line auto manufacturing from industrialized slaughterhouses (Shukin 2009).

Across Canada and the United States, animal slaughter was concentrated in urban areas. Since the 1960s, however, most slaughtering facilities have been moved out of large cities into rural regions, beyond view of the majority of the population. Richard Bulliet (2005) identifies this as emblematic of the "postdomestic era" in slaughterhouse history, as "people are physically and psychologically removed from the animals that produce the products they use, yet most somewhat paradoxically enjoy very close relationships with their pet animals" (quoted in Fitzgerald 2010, 59). The remaining urban slaughterhouses are usually causes of great debate, as people (who may or may not eat the animals killed in those facilities) object to the slaughter trucks, sounds, smell, and overall housing of an abattoir in proximity to their homes. The Quality Meat Packers plant in south Toronto where pigs were slaughtered, a holdover from this earlier era until it was closed in 2014, was a continuous source of controversy, for example. Those pigs will continue to be slaughtered, but they will be taken to one of the other hog slaughterhouses in Ontario. The 750 workers in the facility lost their jobs.

Slaughtering processes have also been further mechanized, and the number and speed at which animals are killed and their carcasses processed and packaged has increased substantially in most operations. As Fitzgerald (2010, 62) explains, "To put this in perspective, in the early 1970s, the fastest line killed 179 cattle an hour; today the fastest kills 400 per hour." Depending on the facility and the animals, the numbers vary. Timothy Pachirat's (2011) study of a cattle slaughtering facility is called *Every Twelve Seconds* for a very deliberate reason. Many plants slaughter and process more than 4.5 million pigs per year; the number for chickens is even higher (Stull and Broadway 2013). Facilities that take in, kill, and process chickens commonly require workers to hang 20 live chickens per minute by the feet at the start of the disassembly line. In combination with industrialized farming facilities, these changes to slaughterhouses are how and why tens of billions of animals can be raised and killed for food annually, even as fewer people farm. Similarly, because of these structural changes, work in agriculture in the global north is less likely to be about your own farm, and increasingly about working for (usually low) wages for an agricultural company doing dirty and dangerous work. In some cases, such as with Tyson Foods in the United States, the same company may own all or many stages of the food production process. Because of corporate consolidation and the strategic purchasing of smaller facilities and companies, a handful of agricultural corporations own much of the meat production and processing system, where work with animals intended to be food occurs.

Labor involving animals intended to be food is not monolithic; there are different ways that the work is organized, and somewhat diverse visions about the process and about animals. Aboriginal peoples have distinct understandings in comparison to white, capitalist, settler cultures, and there are also differences among indigenous nations and communities. The discussion here focuses on dominant, capitalist, non-Aboriginal ideas and practices, particularly in Canada and the United States. It is important to recognize the dominant patterns, as well as the heterogeneity, in order to properly understand this kind of animal work. But what unites all labor with animals intended to be food is that it is always, ultimately, about preparing the animal for death. While the animals are alive, there are differences in what and how work is done. The differences are shaped by the type of farm, its location, and the people involved, yet there are also similarities. Body work and dirty work are widespread, and some care work and communication work is done to feed, water, and identify and deal with health issues. Arguably, there is an instrumentality to many examples of care work in agricultural contexts, as the animals are being kept alive only temporarily and for specific functional reasons. Moreover, many farming organizations explain/justify the storing of chickens, pigs, and cows indoors year round by saying that this provides them with "protection," yet virtually none of these animals are kept alive for more than a few years or months. In other words, the animals are being protected from a few natural predators, but only until they will be killed by humans.

On some farms, animals' health issues are treated, while on others the animals may be sent to slaughter right away while sick or hurt, or killed on the property, with or without a veterinarian's assistance. What happens will depend on the farm, the worker, and the animal, particularly the animal's role in the larger operation. For example, sustaining the life of a cow deemed a successful breeder may be a higher priority than that of an older or younger animal intended to be killed and consumed promptly anyway. But, at the same time, whether a breeding animal is saved will vary and may depend on the species, because, in countries like Canada and the United States, female pigs (sows) are kept by the hundreds or thousands in gestation crates for most of their lives, able to lie down and stand up, but not turn around. A sick or injured sow among thousands may not be deemed worth treating or saving, and she will instead be put onto the slaughter truck. Among the animals unloaded at slaughterhouses, there are regularly sick, injured, and/ or pregnant individuals. Some animals also die en route, particularly if the journey is long.

It is easy for those in cities and/or on the outside to make blanket statements about the country as being only a place of violence against animals, and to homogenize and demonize agricultural workers, whether on farms or in slaughterhouses. Undoubtedly, contemporary industrial agricultural practices are often about mechanized, for-profit suffering. The truth about the workers involved is more complex, however. Not only are there differences among people and labor processes, but those who work with animals in rural communities also have different views and feelings about themselves, their work, and the animals, as well as different reasons for doing the jobs. There have been and continue to be examples of individual and collective actions that reveal kindness, empathy, and the complexities of emotional work with farmed animals intended to become food.

Rhoda M. Wilkie's (2010) ethnographic research on/with farmers, for example, illuminates a more heterogeneous and nuanced picture. Some farmers and farm workers take a detached or utilitarian (in this case, meaning functional) approach to the animals, while others form close emotional bonds and relationships, particularly with certain individuals (see also Hansson and Lagerkvist 2014; Theodossopoulos 2005). Wilkie (2010) argues that animals are placed on a commodity-companion continuum, often shaped by what the animal's role is on the farm. To capture this complex dynamic and recognize that farm workers can see animals as both individuals with personalities and feelings, and a product that earns them income, Wilkie proposes the concept of "sentient commodity." She also enlists Robert Merton's argument that people in these kinds of complicated working contexts experience ambivalence "not because of their idiosyncratic history or their distinctive personality but because the ambivalence is inherent in the social positions they occupy" (Wilkie 2010, 135). Farmers and farm workers earn a living because animals' bodies are turned into products to be bought and sold (i.e., commodified), yet by working with animals, they recognize that animals are sentient beings, not vacant, unfeeling, and unthinking objects (see also Porcher 2011; Porcher and

Schmitt 2010). At the same time, Peter Dickens (1996) argues that industrialized capitalist reorganization of agricultural practices—and therefore labor—has fundamentally changed the way people understand and know nature. As Jocelyne Porcher explains:

> the industrial organization of work, the denial of the intersubjective bond between farmers and animals, and the repression of work rationales that are not economically based have triggered a deterioration, if not a perversion, of the relationship between workers and animals…The relationship with farming animals has never been an easy one and, like human beings, animals suffer from the violence of social and human relations. The fact that some farmers mistreat their animals has been observed for decades. However, the industrialization of work has profoundly changed the nature of violence towards animals, which is no longer individual or limited to small numbers but has become institutionalized, linked to the industrial organization of work. (Porcher 2011, 5)

Put concisely, in farming and all other contexts, emotional work is both shaped by workplace relations, and it shapes them. As Wilkie (2010, 129) argues, "the attitudes, feelings, and behaviors of byre-face [farm] workers cannot be uncoupled from the productive role of both humans and animals in the practical division of labor (e.g., breeding, storing, and finishing) or the socioeconomic context." Different types of work and the specific occupations therein require specific tasks to be done, including emotional labor and/or emotion work. Colter Ellis (2013, 2014) has found similar dynamics on US beef farms, for example, as farmers negotiate the caring/killing paradox by emphasizing differences between individual beings and the products that will be produced, and by actively working to build the emotional skills necessary for managing and negotiating their work and its implications. Ellis and Leslie Irvine (2010) argue that young people are socialized into farming cultures by their family members and through programs like 4H where children and teenagers are taught not only technical skills and knowledge, but also what is intellectually and emotionally desirable, justifiable, and essential for farming livelihoods and communities that depend on animals' deaths. This is not always a totalizing process, however. Of the experiential and relational dynamics on farms, Jocelyne

Porcher writes, "The place of death in farm work means that the farm animal is *almost* a friend. Sometimes, despite the farmer['s emotional management], the animal becomes a real friend, and that is why some amongst them cannot bring themselves to send certain animals to the abattoir, and prefer to keep them to finish their days at home, even if this choice is expensive from an economic point of view" (Porcher 2016 n.p.). Moreover, Porcher (2011) argues that processes of suffering can still spread from animals to people. Her research reveals that although farmers and farm workers employ many emotion management strategies, animals' distress deeply affects some people physically, psychologically, and emotionally, especially women.

Undercover exposés compiled by animal rights organizations reveal that laws are violated in agri-corporate facilities; laws also permit treatment that many people would find unacceptable if they were made aware. In such exposés, the conditions are exposed, but so, too, are examples of human workers beating and/or torturing animals. In some cases, animal cruelty charges are laid because these acts stand out as particularly sadistic, as well as illegal. Employers condemn these acts and often claim ignorance, then promise to reassess their workplace practices. Yet broader questions are always raised about whether these are isolated incidents. People wonder why workers would beat turkeys, pigs, cows whose lives are already shortened and slaughterhouse-bound. Some people question as to why the whole system is not deemed animal abuse.

The lousiness and precariousness of the work no doubt contributes to certain workers' particularly violent acts, and those with options are likely not choosing poverty-wage farm work. "Rarely… do media headlines connect human and animal cruelty (hum/animal) to the exploitation and poverty of living in economically deprived communities" (Renold and Ivinson 2014, 366). Moreover, work-places that hold thousands of largely immobilized animals in bat-tery cages or crates have industrialized cruelty structured right into their very core. Such structures create and perpetuate cultures of deep commodification, devaluation, and suffering, and this shapes how workers therein see animals. Many quit as quickly as they can, unable to tolerate the daily practices and work requirements, and unwilling to reproduce such misery. It is especially those with few options who stay, although clearly certain people are comfortable with the hierarchies, conditions, and workplace requirements. Some of them translate the systematic degradation of animals into individualized eruptions of violence. It is possible to understand how and why it happens, but that does not make it acceptable or the perpetrators inculpable. There is a long history of some people, and particularly certain men, turning their anger at their own exploitation into violence against the women, children, and/or animals in their lives. In these cases, those who feel oppressed, disrespected, and devalued, seek to feel that they have power over something/someone, and thus harm those they deem "below" them. There are also troubling connections between violence that starts with animals, and then expands to children and women, and/or other people (see, e.g., Ascione and Arkow 2000; DeGue 2011; Gullone 2012; Linzey 2009; Tiplady, Walsh, and Phillips 2012).

Yet most workers do not act in such ways. There are many who refuse to exacerbate situations of harm and who do not turn their anger onto others, even in situations of institutionalized devaluation, which say to both humans and animals that they are disposable, the latter more literally so. Those who stay but do not extend workplace relations of domination into individual acts of violence demonstrate the potential of agency in some ways, but simply refusing to be extra violent in systems rooted in domination and disposability is not sufficient on its own. The "whistleblowers" with the courage to speak out publicly warrant recognition, as do those working to change perceptions and patterns in different ways, including behind-the-scenes, or by doing work differently, thereby demonstrating alternatives. For example, although dominant practices in modern dairy production in the global north involve taking calves away from their mothers, usually a day or so after birth, there are a small number of farmers who, out of a desire to balance their need to earn a living with their commitment to the animals, take some milk from the cows to be sold for human consumption, but also allow calves to stay with their mothers. This was done more widely historically and continues to be more prevalent among small holders in the global south.

Overall, the material realities of work do not *determine* how people think and feel, but they help *shape* perceptions of self and others, within and across species. The structure of workplaces creates conditions where particular ideas are easier or more necessary, while alternative interpretations and analyses are discouraged. At the same time, the data reveal that the very same types of workplaces, such as primate laboratories, can be run and experienced differently based on the approaches and resulting practices of the people therein (Arluke and Sanders 1996). In other words, the structure of a workplace is part of the story, yet those involved can affect daily work relations and lives. Put another way, structures matter, but so does agency.

Contextual dimensions like where the farming take place and what other employment opportunities exist in these regions also figure and influence who does what work and why. This is interwoven with the socioeconomic conditions for specific groups of people. Someone may have a deep connection to farming based on their family's past. Someone may also respect the craft of butchering if their family owned a shop and took pride in that skill. However, the overwhelming majority of people do not want to work in today's slaughterhouses, for example, even if the pay is above minimum wage. Ron Davison (2014, n.p.) of the Canadian Meat Council (CMC) argues that the industry relies heavily on temporary foreign workers because Canadians do not work to do the work. "We have tried to recruit Canadians, extensively and constantly. If you go on the job bank, eight CMC members have job opportunities and we aren't getting Canadians to do these jobs…It's particularly a problem for the rural areas, where the plants are located…We don't know more what we can do to recruit Canadians. That's the problem. The whole industry is trying to do it, and we just aren't getting people to [work]." This is not an uncommon refrain espoused by spokespeople for a number of agricultural industries, and its validity has been contested in certain instances. However, given what is involved in slaughterhouse work specifically, it certainly seems possible that recruiting and retaining workers would be a challenge. As a result, poor people from other countries are increasingly being offered temporary employment contracts, often lured not only by the prospect of pay, but also by the (usually unfulfilled) possibility of gaining permanent residency or citizenship in a country like Canada. Of the local workers employed, marginalized social groups are now disproportionately represented in this kind of work, as well, particularly recent immigrants, women, and racialized people (Nibert 2014; Pachirat 2011; Stull and Broadway 2013). In other words, poor people who do not have a lot of options, whether from nearby or elsewhere, are forced by economic necessity to do slaughterhouse work.

Virgil Butler was a slaughterhouse worker who quit and then became an advocate for animals as well as a critic of his former employer and the system it represented. He shared these crucial words: "So many activists say horrible things about the people who toil away in these miserable conditions and do these torturous things to birds, but what they don't realize is that these workers are in a lot of cases just as much victims of the industry as the chickens, and that the reason they do unspeakable things is mostly because of these conditions. Most people think that only monsters could do this job, when in reality, most of them are simply poor people trying to feed their families and have no other options" (Corman 2005). This is the heart-breaking reality, one that speaks volumes about the interconnectedness of people and animals. The literal truth is that it is chickens and other animals who are killed continuously and incessantly at slaughterhouses, but Butler's point about linked devaluation and exploitation is clear. The work in slaughterhouses, in particular, is physically, emotionally, and psychologically difficult, as well as dangerous. The smell, the blood, and all too high a number of animals who do not get killed at the outset and strike at workers due to pain or fall "off" the assembly line and frantically run around in terror are visceral, daily or even hourly reminders that slaughterhouses are industrialized factories of killing and dismemberment where living beings enter but do not leave alive or whole (Pachirat 2011). This is compounded by the serious health and environmental impacts of the work on those inside, as well as on surrounding communities (Broadway 2000; Fitzgerald 2010; Stull and Broadway 2013). Some research also suggests there are increases in crime rates in the surrounding communities after slaughterhouses are opened, including violent crimes (Fitzgerald, Kalof, and Dietz 2009).

Regardless of the pay and even if some protections are afforded through unionization (the likelihood of which varies depending on the context), it is tough to describe slaughter-house work as pleasant, let alone rewarding. Rates of turnover can reach 200 percent per year in slaughterhouses (Stull and Broadway 2013). Similarly, in the industrial livestock operations, which dominate North American agriculture and house thou-sands of hens in battery cages wherein they cannot even spread their wings, or sows in gestation crates unable to turn around, not only are the animals literally constrained, but the likelihood of forming relationships is severely stifled, as well. The emotion work is distinct and complex, as the death of a beloved family's dog requires a particular kind of response, which stands in stark contrast to daily knowledge that every one of the thousands of animals you see not only lives a life of deprivation and is unable to express her/himself, but is also going to have an unceremoni-ous, premature ending. Yet even if humans do reject the institu-tionalized commodification of animals and the exploitation of working class and poor people, and express kindness in whatever way possible, the material reality remains the same. The ends do not change, and the cause of the suffering continues. Ideas and feelings on their own do not change structures. This is an unde-niable and unavoidable reminder of the need for human workers to not only think and feel particular things and to engage in indi-vidual daily acts of compassion, but also to envision—and pur-sue—alternatives and broader, substantive, interspecies changes. It is to these areas of possibility that I return in chapter three and the conclusion.

This chapter has sought to identify and assemble the main, existing threads of research, as well as offer concepts and insights to help us understand the work done with/for animals. Together, these facets create a holistic vision of and approach to human work in multispecies and interspecies contexts, intended to help foster contextualized, nuanced understanding. Neither exclusive exami-nation of the structure of work nor its experience illuminates the whole picture, especially not when multiple species are present. By approaching the personal in its local and larger structures, and considering the political economic along with the experiential, we are best positioned to understand the who, what, when, where, and how, as well as the why.

Each subsection presented warrants more analysis, and the concepts included can be enlisted to help unpack and understand a broad, cross-section of animal work occupations and contexts. Work with/for animals is a process, not only understood in the sense of official labor process approaches, but also as an embodied and lived experience, one that involves whole living beings who think, feel, share, and understand. Animal work is messy, both literally and figuratively, and it is a process of negotiation, one located within the larger conditions and systems established by people, specifically by those empowered with the ability to organize not only work, but also life, its experience, and its end. Work with/for animals is a multifaceted process, always contingent and context-specific, at once structured and unpredictable. This realization prompts us to take the "multi" and "inter" in multispecies and interspecies seriously, in our understanding of work, and in our ideas about what work is and could be. Notably, most work with/for animals is also work with/for people. Overall, we are prompted to recognize not only diversity, but also commonalities and connections, within and among individuals and species.

2

The Work Done By Animals: Identifying and Understanding Animals' Work

In a discussion I initiated about horses' work on a listserv for equine researchers, one member asserted that horses may exert physical energy, but that they cannot work, that they cannot "have a Protestant work ethic." The statement certainly betrays a narrow and ethnocentric perception of work, one that would exclude the labor and contributions of most people in the world, in fact. It likely is also reflective of some people's discomfort with the prospect of not only acknowledging but recognizing animals' work—and what doing so might mean for our social and economic relations.

People's ideas about animals' work are complex, overall. In show jumping horse cultures people regularly talk about making a horse work (in the sense of further engaging particular muscle groups or doing a specific exercise/task) or even of the horses as liking and/ or understanding "their job." The latter can have context-specific connotations beyond the tasks of jumping a course quickly and cleanly to win the most prize money and glory for the rider and owner(s). For example, a horse ridden by an amateur is expected to not only respond to the aids applied by the rider that adjust speed, gait, length of stride, and so on, but to actively disregard those directions that the horse deems incorrect, and to instead proceed in a safer manner. Horses ridden by amateur riders are expected to take care of the people on their backs as part of their jobs, yet show jumping horses are not considered "workers" (Coulter 2014; Thompson and Birke 2014). They may, however, be considered athletes, and it is not uncommon for both lay people and scholars to conceptually differentiate such animals from those who are "working" (Fennell 2012; Kaushik 1999; Nance 2014). Dogs provide another illustration of the social construction of these demarcations. "Working dogs" are usually identified specifically as those doing police, guide, search and rescue, service, or herding work. In this chapter, however, I propose an expanded and multifaceted way of identifying and reflecting on animals' work, and on the connections and distinctions between different kinds of labor.

While writing and teaching about multispecies and interspecies labor, I have found that even the idea that animals work elicits either fascination or resistance. Some people become intrigued, and see this fact as obvious yet underacknowledged and underexplored, then their minds begin to actively reflect on the different ways animals work. The popularity of the Canadian-made television program "Dogs With Jobs" in the early 2000s is testament to many people's interest in and curiosity about animals' work. Yet for others, there is cynicism, disbelief, and/or hostility, each dimension likely stemming from particular intellectual or political qualms. Nevertheless, animals work in virtually every community on earth and in diverse ways therein. This should neither be denied nor ignored. Rather, the specifics of animals' work warrant careful examination. Moreover, if committed to building a genuinely multispecies study of labor, we also need to develop the conceptual tools and frameworks for thoroughly understanding and analyzing animals' work.

As part of a broader push for post-humanist thinking, Donna J. Haraway (2008) has suggested that there is value in developing ways of thinking about animals' work that do not rest on humanist intellectual frameworks or concepts. Post-humanism has different meanings depending on the scholar or area of scholarship, but, overall, does not mean after humans but rather beyond humans; it is about decentering humans as the primary focus and subjects of inquiry. This chapter is motivated by the premise and promise of such a project, and by the goals of a more inclusive approach to both animal studies and work and labor studies. My concept of animal work as a whole has been deliberately developed to take seriously the "multi" in multispecies. At the same time, there is intellectual and political value in both focusing on animals, and in seeing how their lives (and deaths) are implicated in anthropogenic and anthropocentric structures, and interspecies relationships and labor processes. In other words, animal-centric inquiry, multispecies work, and interspecies labor all warrant analysis, and the on-the-ground realities demand that we do so in a politically engaged way. Thus Lynda Birke's (2009, 1) query is worth repeating and remembering: "What's in it for the animals?" This is an analytical and a conceptual chapter, but it is connected to larger political and ethical questions not only about the goals and effects of scholarship and intellectual labor, but also about the problems and promise of work more generally.

Accordingly, I have opted to pursue a few intellectual routes in this chapter. I engage with and enlist a number of existing humanist concepts and frameworks. These are the primary ways of thinking about work, and discounting their potential usefulness is both ahistorical and unnecessary. I am not persuaded of the need to entirely reinvent the wheel or throw the puppy out with the bath water when it comes to understanding animals' labor. The existing scholarly vocabulary is helpful and offers a number of fruitful concepts and lenses. However, notably, some overly narrow and exclusionary frameworks have informed dominant ways of thinking about people's labor. Formal jobs have been given greater attention than unpaid or informal work. Similarly, employment sectors that are more male-dominated have gained the most scholarly and political attention, particularly those that are for-profit. This way of thinking about and framing work has undoubtedly contributed to both popular and academic perceptions of what constitutes animals' work. I do draw from theories that are dominant in the labor studies field, but given the specifics of animals' work, there is also an irrefutable need to emphasize gendered and feminist analytical frameworks and approaches. Thus I draw from the specifics and spirit of a feminist political economy lens throughout this chapter and book.

At the same time, because animals' work is simultaneously similar to and distinct from humans' work, there is a need to push past existing theories and to develop concepts that properly recognize, capture, and explicate the complexities of nonhuman labor.

Consequently, I hope we can build from but also beyond existing thought. In this chapter, I assemble and interweave existing theory and research, enlist and assemble some underconsidered approaches, and propose ideas, frameworks, and questions to propel the discussion. As a result, this book as a whole, and this chapter in particular, contribute to what will hopefully be a broader, collaborative project of discovery, and a larger political conversation about the interpersonal, socioeconomic, and political implications of not only understanding, but of recognizing animals' work.

Beasts of Burden? A Labor of Love?

Akin to examining the work done with/for animals, even trying to simply identify the breadth of work done by animals today is a major task, particularly if thinking cross-culturally and globally. The first working animals that come to mind are likely police and service or guide dogs, and horses pulling carts, wagons, or carriages. These animals are tasked with visible, physical tasks clearly identifiable as "work," and/or are ascribed with a sense of formality or authority. The latter is achieved through titles, accreditation (e.g., recognized service and therapy dogs are usually certified and have papers stating so), their close association with easily identifiable and widely recognized human workers (like police officers), and/or specific visible markers like harnesses, collars, or blankets.

More broadly, the descriptive categories that apply to work with/for animals also apply to animals' work. Many sectors involve both interspecies and multispecies work, as humans work alongside animals who are also working, in different and similar ways. Today, animals work in transportation, "resource development" (such as mining), service, law enforcement, military, health care, education, entertainment, sporting, tourism, and/or agricultural sectors. Animals, especially equids (horses, donkeys, mules, jennets) and oxen, as well as camels, llamas, elephants, among others, haul people and goods in a range of places, especially in poor communities across the cities and rural spaces of the global south. The precise number of draft animals working today is difficult to determine, and estimates range from a few hundred million to around one billion. In some countries, the number continues to increase. In

1999, half of the global human population was "heavily dependent" on draught animals' work (Kaushik 1999). As noted, the Food and Agriculture Organization (FAO) (2009, 2011) estimates that close to one billion, or 70 percent of the world's most impoverished people, rely on livestock for their livelihoods today.

These animals' roles are much more diverse than the term draught animal suggests. A report called "Invisible Helpers" by the international nongovernmental organization, The Brooke (2014), exposes the multifaceted nature of equids' laboring contributions. Researchers found that animals often work in fields and on roads, or, in both production and distribution. Many animals help poor women with household chores such as fetching water and taking children to school. Animals' manure is also used as fertilizer. Indeed, animals' work in the broad sector of agriculture can mean very different things. A sheepherding dog, sheep-guarding donkey, and plough-pulling ox are all working in agriculture. Similarly, David A. Fennell (2012) points out that in tourism, animals are used for pulling strength, speed (e.g., dog sledding), riding, pack/carrying labor, among other forms of work. Put another way, animals are involved in different occupations and types of labor even within a single sector or work site.

The individual use of service, guide, or assistance animals in countries like Canada is growing although somewhat unevenly because of the costs associated with their training and/or with obtaining and sustaining one (food, veterinary care, etc.). Service animals are tasked with navigating complex terrain to safely guide people to work and school, and through the demands of daily life. Animals carry out essential daily tasks (from turning lights on and off, to opening and closing drawers and cupboards, to helping with laundry, and beyond), and offer various kinds of health care and therapeutic service, including comfort, relief from serious anxiety, panic attacks, and posttraumatic stress, and/or warnings of seizures. This job means real-time predictive or responsive responses, and round-the-clock involvement in serving someone's needs. Dogs in particular are used for this kind of direct, interpersonal work, but so, too, are rats, miniature horses, monkeys, among other animals. Thus, as with human workers in personal support and other interactive health professions, these animals' work is about both service and health care. Clearly the descriptive categories outlined begin to tell the story but do not thoroughly capture the depth and breadth of the work done by animals in and across these sectors. As noted in chapter 1, the work done with/for animals is not yet well understood; animals' work is even less studied. Substantial intellectual labor, reflexivity, and humility is needed to right (and write about) this wrong.

Building on anthropological approaches to livelihoods and feminist political economy, I propose three conceptual categories to help foster organized understanding of the breadth of animals' work (figure 2.1). At the broadest level, animals, especially in the wild, engage in subsistence work. This work is shaped by people's settlements, infrastructure (like roads and hydro lines), and certain leisure or livelihood-based pursuits (such as hunting and trapping), as well as by humans' influence on the environment and climate (droughts, the removal of forests for logging or agriculture, etc.). Yet subsistence is the type of animal work that involves the least direct contact and interactions with humans and people have not asked for or mandated this work. There is little evidence to suggest that these animals should explicitly be called "workers," but this kind of subsistence labor should nevertheless be acknowledged as such; we certainly identify people's subsistence work as work. If interested in identifying the breadth of work done by animals, subsistence work in the wild ought to be noted and studied to some degree.

The second type involves voluntary work, usually done for humans. The most widespread and common example of this is the care and protective work animals provide in homes. Individual animals do not usually choose to be in specific homes, but they are

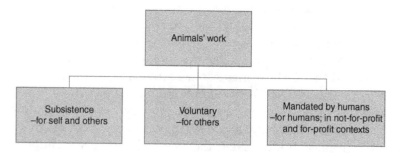

Figure 2.1   Animals' Work

able to exercise some control over the degree to which they provide protection and/or care work. There are also animals who voluntarily assist other animals, even across species lines, particularly those who are physically disabled.

The third category is the work mandated by humans. This involves formal work given to animals and includes a broad cross-section of tasks, assignments, and occupations. Animals are able to exercise differing degrees of voluntary, self-initiated and/or self-controlled activity within these occupations, but the occupations overall are chosen by people. Each of these three types of work influences animals' well-being, but the third category is most central and significant, and warrants the most analysis. In these cases, animals not only do work, but they can most clearly be considered workers. The identity of "worker" is contested and differently applied by/to human workers, thus warrants more discussion in an interspecies framework; in this chapter, I focus primarily on work done by animals. Work done by animals is clearly linked to notions of animals as workers, but the two ideas are not universally connected or automatic. Within the third category, work done by animals can be subdivided into descriptive categories (e.g., service, law enforcement, tourism, etc.), further unpacked through the enlistment of labor process and other theoretical concepts, and analyzed based on evaluative criteria, particularly about animals' experiences. Accordingly, these categories provide only a starting point.

Every Animal Mother Is a Working Mother, Too: Social and Ecosocial Reproduction

Twenty-five years ago, economist Marilyn Waring (1990) fundamentally challenged conventional calculations of worth and value by pointing out that both unpaid work in homes and nature's processes are not counted.[1] In other words, the labor performed in homes by both women and men but especially by the former, makes all economic activity outside the home possible, yet it is not formally recognized. Similarly, nature's contributions are not valued and often not even acknowledged. For example, a forest performs vital functions by processing carbon dioxide and emitting oxygen, but this life-saving and life-giving role is rarely noted. Only if people visit forests and generate tourist revenue, or if forests are cut down, do these trees figure in mainstream economic measures. Both these insights are a helpful starting point for examining animals' subsistence and voluntary work.

Every wild animal is responsible for sustaining him- or herself, alone, or through shared labor with others. Daily life for animals involves the rigorous and multifaceted challenges of survival—from avoiding predators and threats (both natural and human, including vehicles), to finding food and water sources in all seasons, regardless of the weather. In other words, animals do subsistence work for themselves and often for/with others. This is life-sustaining work that living beings have always pursued, long before people began studying such processes. But people's impact on the natural environment, including the clearing of land, the erection of buildings, cities, and dams, the laying of roads, highways, and train tracks, and broader changes to the climate and environment (such as dried-up or poisoned rivers, droughts, floods, and so on) have fundamentally altered animals' historic patterns and strategies, and made contemporary survival even more challenging.

Feminist political economists have developed ways of highlighting and thinking about people's unpaid work which I argue are applicable across species lines, although so far they have not been given much attention by human-animal scholars (see, e.g., Bakker 2007; Bakker and Silvie 2012; Brodie 1995; Folbre 1994; Luxton 2009; Luxton and Bezanson 2006). First, social reproduction is a concept that reflects the fact that unpaid work is both a set of tasks and a process (Bezanson 2006). This means that specific unpaid tasks are continuously required: cooking, cleaning, laundry, shopping, and so on. The overall, cumulative effect of these tasks is the larger social process of reproducing people, of ensuring present and future generations of workers. Social reproductive work is also about teaching language; ways of acting, relating, and thinking; emotional skills, and the like. Whole people are continuously sustained, as living beings and as social actors. While social reproductive work can be done for pay in houses, child care centers, and schools, much of it is always done without pay in homes. People are fed, clothed, bathed, healed, nurtured, organized, taught, entertained, scolded, empowered. In some places, there is a more equitable distribution of this work among genders, but the bulk of social reproductive labor continues to be done by women (namely mothers) across contexts.

In homes and families, animals also contribute to people's social reproductive labor through their provisioning of care work, whether guarding, monitoring, or comforting children (and adults), and/or helping transport people and daily necessities like food and water. Medina Hussen from Ethiopia explains: "If there is a donkey in the house, the mother carries her child on her back and lets the donkey carry other stuff such as water or crops. But if she doesn't have a donkey she has to leave the child behind at home even when there is no one to look after the baby, as she has to carry the load herself. So donkeys have a huge contribution in caring for babies" (The Brooke 2014, 35). There is also a measure of interspecies reciprocity here as children are taught how to perform unpaid work themselves, including by helping to care for animals, and, of course, animals who live in people's homes are also the recipients and beneficiaries of human's care work. People who have animals in their homes feed, water, exercise, entertain, and nurture them and this means humans engage in unpaid care, emotional, body, and dirty work for animals. These dynamics are powerfully evident among the human-animal families without stable housing and/or who live on the streets, where animals, and especially dogs, provide protection, warmth, companionship, motivation, and emotional support (Irvine 2010, 2013; Labreque and Walsh 2011; Lem et al 2013). When women are being abused, their animals can provide social support that is life-sustaining (Fitzgerald 2007).

Undoubtedly, the emotional and care work animals do in families providing joy, comfort, and compassion is immeasurable. Animals continuously assess the people with whom they live, physically, intellectually, and emotionally, and then proactively and/or responsively provide care of various kinds. Because this is ubiquitous, some people may take it for granted, or value animals' roles, but not see this as work. Even when talking about human-human care work, particularly when done without pay in homes for reasons of love and/or familial responsibility, some people are resistant to the idea of considering this work (England and Folbre 1999). However, care work is work, and recognizing that caring involves physical, intellectual, and emotional labor does not mean it is not or cannot be motivated by love, or that it cannot be viewed in other ways, simultaneously.

At the same time, animal mothers also engage in their own social reproductive work around the world. In fact, people directly benefit from the reproductive labor of animals in the wild. For example, as bees collect nectar to feed their young, they pollinate nearly 70 percent of all flowering plants, which allows those plants to reproduce. Insect pollination is involved in and/or integral to sustaining over 30 percent of the foods and beverages consumed by people (Centre for Urban Ecology n.d.). Depending on the specific animals in question, this kind of work may or may not be ensuring future generations of workers for a capitalist system as human social reproductive labor does. In any event, individual families and whole species rely on the continued existence of their kind, whether they are actively involved in the formal economy or not. Thus wild animals' work is essential to the reproduction of ecosystems. As a result, I propose that we can also think of them as engaging in ecosocial reproduction.

Animals reproduce in the sense of conception and birth, but they also engage in reproductive labor once babies are born. As I write this chapter, outside there is a robin who spends the majority of every day sitting on the nest that she and her mate have built, which houses the eggs she laid. She is dedicated and dutiful, and spends hours on those eggs, monitoring her surroundings, and assessing people, dogs, cats, other birds, and vehicles for potential threats. Once the eggs are hatched, she will continue to monitor her offspring, and will ensure they have appropriate and adequate food. She will keep the nest clean by removing waste. If threats approach, she and her mate will call out specific alarms, and may attempt to challenge the attacker. When it is time for the fledglings to fly, she will model the desired behaviour and may peck or push at a youngster in encouragement. The young robins will continue to follow their parents for a couple of weeks after learning to fly and may beg for food, until they learn to become self-sufficient. What else the robins may be doing to ensure the health and safety of their dependents, humans, even biologists and ornithologists, may or may not fully understand. Only recently, for example, have researchers learned that cows have different types of calls for their offspring (de la Torre et al. 2015).

Human social reproductive labor is largely unrecognized; both domestic and wild animals' contributions are even less acknowledged. One possible area of exception is in acute or emergency situations, such as if a person has an accident and the animal notifies someone, and these kinds of "hero" animals are increasingly celebrated. Accordingly, the idea of social reproduction is central to both animals' own subsistence work, and animals' voluntary work in homes. Different kinds of care-giving labor are also integral and when taking stock of the work done by animals in their own families and in multispecies families in human homes; the concepts of social reproduction and care work are complementary. Some animals are prevented from engaging in social reproductive and care work, however, and these contexts will be revisited shortly.

The Birds, the Bees, and Marx

Work performed by animals that is mandated by humans is broad and multifaceted. As with animals' subsistence and voluntary labor, I suggest that there is a need to enlist but also expand anthropocentric notions and theoretical concepts. The formal study of work and labor stems primarily from early sociological thinkers and especially the texts of Karl Marx, thus his work is a prudent starting place. Marx's writings were analytical and/or political, with some texts decidedly agitational, and others more diagnostic. Living in mid-nineteenth-century Europe, a time of great economic and political debate and mobilization, he wrote detailed dissections and theses on capitalism, class, and labor as a social process. Put another way, for Marx, labor is inextricably connected to the organization of both the economy and society, thus structures people's social positions and shapes their experiences and understandings. In particular, Marx highlighted people's relationship to production: a few people own productive infrastructure (members of the capitalist class), while others have only their labor power, which they sell in exchange for a wage (members of the working class). Marx's writings are expansive and involve far more material than can be synthesized here. What is most pertinent is that given his emphases and the largely Eurocentric study of work that was subsequently developed, much scholarship on labor reproduces quite specific images of what work is and of who is a worker. Those involved in literal production—that is, factories—have widely been constructed as the standard and central "workers" because they produce "value" in a capitalist context, and are positioned in economic locations deemed politically, strategically, and tactically essential by those interested in challenging and replacing capitalism.

Clearly, even in Marx's time, such framings of work and workers did not emphasize the breadth of the work that was going on or all the people involved. For example, those in educational, health, or other kinds of service work (such as retail) are not emphasized. Those working in rural regions pursuing subsistence or other kinds of economic relations are not a central or significant focus either. And without question, unpaid work being performed in homes everywhere was not seriously analyzed as labor. More recently, the work and labor studies lens has been expanded. Sociologists and anthropologists of work have paid the most attention to service work and rural contexts, cross-culturally. Some have enlisted elements of Marx's frameworks and others have not. More scholars now have a broader view of many forms of "value" produced by workers of different kinds. Yet service, rural, subsistence, and unpaid work (among other types) continue to be relatively understudied in a number of labor studies circles, especially in comparison to industrial and manufacturing workers, and particularly those who are in unions. This is despite the fact that the majority of workers on the planet work in service or agriculture. Thus, in part because it is men who primarily did and do industrial work, and in part because of historical patterns of inclusion and exclusion in universities more broadly, dominant trends in labor thought and action have tended to be androcentric (male-centric). Moreover, the study of work, like most areas of scholarly inquiry in the social sciences (and beyond), has undoubtedly also been anthropocentric (human-centric).

The suggestion that a singular text or thinker offers complete, infallible, or prescient vision is, of course, incorrect. Laudably, most contemporary writers are not suggesting Marx be read as an omnipotent fortune teller, or that his work be seen as a sort of bible applicable to every issue and context. Moreover, when it comes to thinking about animals' work specifically, Marx may not be the first theorist who comes to mind given the limited and not especially helpful treatment he gives the subject. Yet because of Marx's influence over understandings of labor processes, capitalism, and progressive politics more broadly, his work and ideas have shaped the scholarly terrain and warrant further discussion here.

Marx (1978, 344) wrote occasionally of animals and their work, but did so primarily to highlight what makes human labor exceptional. "We are not now dealing with those primitive instinctive forms of labor that remind us of the mere animal" he writes. "A spider conducts operations that resemble those of a weaver, and a bee puts to shame many an architect in the construction of her cells. But what distinguishes the worst architect from the best of bees is this, that the architect raises his [sic] structure in imagination before he erects it in reality." People transform nature through their work and do so consciously, Marx argues, in contrast to animals:

> The animal is immediately one with its life activity. It does not distinguish itself from it. It is *its life activity*. Man [sic throughout] makes his life activity itself the object of his will and of his consciousness. He has conscious life activity…Admittedly animals also produce. They build themselves nests, dwellings, like the bees, beavers, ants, etc. But an animal only produces what it immediately needs for itself or its young. It produces one-sidedly, whilst man produces universally. It produces only under the dominion of immediate physical need, whilst man produces even when he is free from physical need and only truly produces in freedom therefrom. An animal produces only itself, whilst man reproduces the whole of nature. An animal's product belongs immediately to its physical body, whilst man freely confronts his product. An animal forms only in accordance with the standard and the need of the species to which it belongs, whilst man knows how to produce in accordance with the standard of every species, and knows how to apply everywhere the inherent standard to the object. (1978, 76)

It is clear that Marx did not think about the horses visible outside his window hauling people, products, the mail, and emergency vehicles like fire trucks when he wrote this section. Marx's focus here is only on very specific tasks done by some animals in the wild (not animals' subsistence work in a broader sense), despite the widespread presence of animals working in human-created workplaces and spaces, in cities, rural regions, and on farms. Marx (n.d., n.p.) references horses' work elsewhere in this way: "Of all the great motors handed down from the manufacturing period, horse-power is the worst, partly because a horse has a head of his [sic] own, partly because he is costly, and the extent to which he is applicable in factories is very restricted. Nevertheless the horse was extensively used during the infancy of modern industry. This is proved, as well by the complaints of contemporary agriculturists, as by the term 'horse-power,' which has survived to this day as an expression for mechanical force." In other words, the domesticated animals most visible in a laboring capacity do not figure in Marx's central discussions of animals' work, but rather are discussed as merely a "motor," and one inferior to inanimate technologies, in part because the animal has an active mind.

So there is a bit of unevenness evident. It would appear that, for Marx, animals can think for themselves (literally, "a head of his own") when working a human-constructed context, and cause complications (which sounds remarkably like resistance, a cherished process in his thinking more broadly, when applied to people of the working class). At the same time, the limitations of animals' minds are central to his argument about what makes human labor decidedly different and superior. This tension is, in many ways, a foreshadowing of contradictions evident today in how a range of people view animals and their work. They may recognize that individual animals work and even express agency, but they are still not prepared to extend their analyses or politics to animals. However, if cognition, understanding, and agency form the wall used to separate humans from animals, people today do not have the excuse of living in a time period when animals' minds were not studied and certainly not recognized or understood.

Select interspecies scholars have revisited Marx's work in more detail and engaged critically and/or constructively with his explicit writings on animals and labor, finding differing levels and dimensions of contemporary applicability (see, in particular, Benton 1993; Clark 2014; Ingold 1993; Murray 2011). As outlined in chapter 1, labor process theory has been used by a handful of scholars to think about humans' work in interspecies relationships, but has only rarely been applied specifically to animals' work itself. As noted, the core elements of Marx's labor process theory are purposeful activity, objects of work, and instruments of work, the latter being akin to a tool between the worker and the object of their labor. Arguably, donkeys' work in the mines in countries like Pakistan offers a good starting place for examining whether these three elements are applicable to animals because the tasks performed are clearly visible as "work." In these spaces, donkeys are tasked with entering the mines at least 20 times a day, where the sacks they carry are filled with coal weighing about 40 pounds. Next, the donkeys are guided to the surface where the bags are unloaded. Sara Farid (2014, n.p.) writes, the donkeys "then obediently turn and walk again towards the black hole. The workers have made a choice to be down here, I think, even if it's a bad choice made by poor people with few options. The donkeys haven't chosen this life, but nevertheless they trudge trustingly up and down the tunnels, wounds on their backs and faces covered with coal dust." Many readers will see clearly some or various possible examples of purposeful activity (moving in and out of the mine shaft), objects of work (coal), and instruments of work (bags) involved in the donkeys' contributions. Some may wonder about what role donkeys' bodies play in the process, and whether their bodies are objects and/or instruments, a dimension to which I will return shortly.

Overall, many readers will also feel comfortable applying the labor process elements to a host of other examples of animals' work. Yet as Jonathan Clark (2014) and Jocelyne Porcher (2016) point out, some of the debate and controversy about animals' labor still rests on ideas and assumptions about awareness, foresight, and intention—the purposeful activity dimension of the labor process. Arguably, certain forms of human labor and many specific, tedious, monotonous tasks required as part of waged employment, especially those jobs characterized as "low skill," involve minimal deliberate intent. Some psychological or biological scholars would suggest that many tasks or processes that can be called work are, in fact, at least shaped by instinct or habit, as well. More to the point, the bourgeoning collection of research in cognitive ethology, a scholarly field within which researchers examine animals' minds and emotions, makes it clear that animals of all shapes and sizes have complex and rich inner-worlds, and that people are only beginning to scratch the surface when it comes to understanding most other species (see, e.g., Allen and Bekoff 1999; Bekoff 2008; Gould and Gould 2007). In other words, today we cannot confidently state that beavers do not preplan their dams, or that donkeys do not understand, to some degree, what they are to do in mine shafts. Interestingly, Jonathan Clark (2014) and Tim Ingold (1983) note that Lewis Henry Morgan, writing in 1868 around the same time as Marx, argued specifically that beavers do consciously engage and assess their dams. Moreover, as Jason C. Hribal (2007) emphasizes (and Marx noted), animals also regularly engage with the labor process through their lack of cooperation and/or their active defiance. In the contemporary context, what is increasingly clear is that we cannot generalize about animals' intentions in a range of work contexts, whether humans are present or not. We also cannot simply dismiss animals' roles as passive or see animals as mere instruments for humans' work. As Tim Ingold (1983, 88) writes, "the domestic animal in the service of [hu]man[s] constitutes labor itself rather than its instrument, and hence that the relationship between [hu]man and animal is in this case not a technical but a social one."

Although Marx's writing about animals' work is antiquated, some scholars have argued that what Marx's work does offer human-animal scholarship is a broader set of ideas or diagnostic tools that can be taken up in a contemporary context as part of assessing the historical, multispecies trajectories of capitalist development and expansion (Benton 1993; Kowalczyk 2014; Murray 2011; Perlo 2002; Wilde 2000).[2] Certain animal rights scholars have also used Marxist ideas along with other radical (in the true sense of the word meaning to identify root causes) theories to critique contemporary political economic relations and advocate for an intersectional, emancipatory politics for humans and animals (see, e.g., Hribal 2003, 2010; Nibert 2013; Torres 2007). As Agnieska Kowalczyk (2014) and Harry Cleaver (2000), among others, point out, Marx's work can be read ideologically or strategically, and he can certainly be read theoretically and analytically, as well, even by those who do not agree with his political proposals. Contemporary analysts may enlist specific elements of his writing, combine them with other concepts and frameworks, or be inspired by particular emphases or the overall spirit of a work. These strategies seem more prudent than trying to make a singular theorist's ideas from 150 years ago fit and explain every context and situation today. Barbara Noske (1989, 1997) has enlisted some of Marx's ideas about human labor to develop her own interesting framework for thinking about animals in agriculture, for example, and her work will be considered in greater detail below.

Overall, there is no convincing reason why labor process approaches and certain neo-Marxist concepts cannot be applied to animals, even if Marx's own comments on animals' work are not particularly helpful. Similarly, for those interested in emphasizing a materialist approach to animals' work—that is, what animals materially do—labor process theory applies. However, a focus on the three elements of the labor process only offers a partial understanding of animals' work and provides more of an entry point or springboard. Even in considering donkeys in coal mines, examination of the three elements does not fully capture the breadth of the tasks, relationships, experiences, or contributions involved, or how animals' work fits into larger socioeconomic and cultural structures. Indeed, in their study of human workers, few contemporary labor process scholars focus exclusively on the three elements. As noted in chapter 1, the term "labor process" refers both to the specific elements being studied, and to a broader intellectual approach for understanding, analyzing, and contextualizing work and workers, one that examines how work is organized and experienced. Concepts from chapter 1 such as body work, care work, communication work, and emotional work thus also figure in animals' work, and are helpful to our understanding of it. Moreover, by interweaving different bodies of feminist thinking that further expand how work and workers are conceptualized and valued, a broader, multifaceted approach to animals' work becomes clear, one that is nuanced and contextualized.

Webs of Animals' Work

The core concepts proposed in chapter 1 apply in different and similar ways to animals' work and help unpack the work they are doing for people. Dirty work—the tasks or whole occupations that are considered unpleasant and/or undesirable—is undoubtedly done by animals. Arguably, some of the dirtiest work is done by animals or by humans along with animals. For example, the dangerous and exhausting work done by donkeys pulling coal from mines is undoubtedly dirty work for donkeys and people alike. Removing loads of bricks from kilns in scorching heat is difficult for both human and animal workers, but even more so for the donkeys or mules physically tasked with hauling work.

Animals may or may not always deem the same things unpleasant or undesirable that people do, however. A search and rescue or police dog tasked with sniffing through a garbage dump or a muddy forest in the rain in search of a missing person or human remains will likely feel differently about the process than humans do. Similarly, in these instances, if a dog's work leads to the discovery of human remains, the dog does not fully grasp the emotional implications of this fact for the family, community, and human workers who have been part of the search. Rather, she or he is provided with verbal and physical praise, and the toy that serves as reward for successful work, thus feels joy. However, the dog will also read and respond to the moods of the people with whom she or he works, and despite rewarding the dog, the humans will reveal a more complex set of emotions in ways that the dog will recognize (Warren 2013).

As such, communication work also applies to animals' work. In chapter 1, I elucidated how, in interspecies workplaces, human workers develop an ability to understand and communicate with animals; animals do the same. The concept of communication work prompts us to recognize an interspecies dynamic and the creation of a shared language, although words may not be central or are likely only one communicative component used. Many examples of animals' work require effective deliverance and receipt of communication to convey and share meaning. A police dog has to learn to read and understand specific commands being used to pass along requests and requirements, for example. These may be verbal, gestural, or some combination. There are also more intricate and minute signals that the most engaged police dogs learn to detect. These dogs continuously monitor their human partners, other people, and surrounding environmental factors, thus engage their eyes, especially their noses, and other senses. The dogs assess body language, changes in demeanor betrayed by heightened breath and heart rates, and so forth. Police dogs also are tasked with communicating clearly to people, and the dogs have different "alerts," which involve sound and/or body positioning, and that vary depending on the situation and individual dog. This process means animal and humans work independently, as well as in tandem, and again this clearly illustrates both interspecies and multispecies communication work.

In a similar vein, dogs, too, must learn to control and harness their instincts and feelings, illustrating what could aptly be called emotion work and emotional labor. A police dog learns to ignore food, cats, and most distractions, while simultaneously focusing on the environmental factors most central to the task at hand. In other words, all elements in a space are assessed, and the dog determines those that are relevant, those to be ignored, and how to act accordingly. Regardless of whether the dog might be feeling afraid, enthusiastic, joyful, angry, or any number of other emotions, they are expected to be "professional" and controlled. In certain instances, these kinds of working dogs are tasked with appearing predatory and/or being intimidating. While dogs, especially the German and Belgian Shepherds and Malinois most often used by police and military forces, are capable of being aggressive, to suggest that this is simply allowing the dog to reveal his or her true nature is incorrect. These same dogs can play happily with people, children, and other dogs when not on duty, and spend hours in a calm, measured state while at work, before being asked to instantaneously become assertive. Thus, these dogs are expected to engage in the internal emotion work needed to successfully perform the external emotional labor requirements of their jobs (like their human partners). These kinds of intricate, nuanced processes are interwoven with the complexities of police dogs' labor, and prompt us to recognize what other kinds of work are involved under the umbrella of "law enforcement" work.

Without question, the concept of body work also applies. Body work, introduced in chapter 1 and defined succinctly as work done with and/or to bodies, is clearly involved in physical tasks like bite-and-hold, which tasks a police dog with, not surprisingly, using her or his teeth to hold the body of a suspect. Arguably, when thinking about animal work, the concept of body work can also be expanded to think about how animals' bodies and physical abilities are used for work. As Rachel Lara Cohen, Kate Hardy, Teela Sanders, and

Carol Wolkowitz (2013, 4) note of human work, all labor involves bodies; the concept of body work means seeing bodies as "materials of production." Therefore, when thinking about animals, body work can mean work done with and to bodies in a more complex sense. Arguably, for many animals, their bodies or parts thereof are both part of them, and essential instruments (or tools) of/for labor. For example, the strong smelling abilities of dogs have been enlisted to search for and detect live people (lost or fleeing), human remains, explosives, weapons, drugs, and other animals. Dogs have been put to work using their noses in rural and urban communities, at borders, airports, and ports, in war zones, in conservation areas and parks, and even on water. Thus, here animals' bodies are being used by dogs directly, and by people who are tasking the dogs with specific work, simultaneously. Rats, too, are tasked with detection work, particularly in eastern Africa. Rats, who are physically light and thus not likely to set off an explosion, are trained by humans working for the organization APOPO, to detect buried land mines and then display an "alert." Human workers are then able to mark the locations and safely detonate from afar, clearing massive tracks of land and making them accessible for local people. Both dogs' and rats' abilities to detect diseases, from tuberculosis to cancer, are actively being researched with promising early results (see, e.g., Cornu et al. 2011; Ehmann et al. 2012: Mgode et al. 2012; Roine et al. 2014).

When body work starts and stops for animals is difficult to determine and something that warrants more reflection and study. Animals' bodies play a central role in virtually all work they do. Their bodies can be physical instruments of power, speed, assessment, and tracking. Their bodies can also be fetishized; the mere presence of an animal in a performance setting, for example, provides wonder and intrigue for human audiences, and "elite" horses are also infused with luxurious symbolic capital (Coulter 2014a; Nance 2013b). In other instances, like therapeutic contexts, it is the process of people touching animals' bodies that provides comfort and healing.

So, is a horse pulling a cart, thus providing the body and strength without which the cart would not move, engaged in body work? Is an elephant in a circus required to stand on her hind legs doing a kind of body work? Is a therapy dog who comforts a panicking child through his calming presence and by allowing himself to be patted engaged in body work of a kind? Applying the concept of body work to animals may lead to a never-ending list. If it does, this may not be helpful; alternatively, it may be a powerful way of recognizing the thoroughly embodied nature of so much of animals' labor. Animals' bodies also figure in other complex ways in labor processes that will be further examined below.

When examining the work done by animals, communication work, emotional work, body work, and sometimes even dirty work intersect as part of care work (as is true for their human coworkers). As noted above, animals do care work for their own offspring as part of subsistence work, and voluntarily for people. Animals also are tasked with doing care work for people. Therapeutic and/ or educational programs for individuals and groups take various shapes and require different things from the animals involved (see, e.g., DeMello 2012; Fine 2010; Kamioka et al. 2014). These programs can be seen as part of a broader interface of nature and health care called green care (Berget and Braastad 2011; Berget et al. 2012; Sempik, Hine, and Wilcox 2010; Westlund 2014). For example, equine-assisted learning or therapy (the latter is also called hippotherapy) is used in many countries to help treat addictions, depression, posttraumatic stress, eating disorders, and various other mental health concerns. Tasks can include grooming the horses, directing the horse through a series of steps (from the ground or while on the horse), learning horses' own ways of communicating, among others (see, e.g., Burgon 2011; Cantin and Marshall-Lucette 2011; Carlsson, Nilsson, and Traeen 2014; Dabelko-Schoeny et al. 2014; Pendry, Smith, and Roeter 2014; Westlund 2014). Not every horse is suitable for such programs, and those involved are not only required to be calm and cooperative at the outset, but to maintain patience throughout each session and interaction. Such programs are noteworthy for a number of reasons, not the least of which is because horses are sensitive flight animals who read people's underlying moods and feelings, which are revealed (or betrayed) by their bodies through increased heart rates, and so forth. Thus, while the people involved may be anxious, especially initially, the horses are required to maintain a sense of calmness, rather than feed off the energy of the nervous person; that takes skill and is work, and it provides another example of animals' emotion work.

Service dogs provide round-the-clock guidance and assistance to people with physical and developmental disabilities and are entrusted with immense responsibility; the animals help to transform the daunting and even the seemingly impossible into the achievable. A number of programs pair children or youth with therapy dogs or cats for a shorter period of time, allowing the chance for reluctant and struggling readers to practice without judgment. Animals are also used as encouragement and partners in physical therapy, providing motivation, support, and reassurance. In all of these cases, the animals provide comfort, serving as part teacher, part facilitator, part friend, and supportive audience. Animals are also starting to be used in certain courtrooms as sources of calm and comfort for children asked to testify in stressful cases, particularly those involving sexual abuse and family members. Indeed, the healing power of animals' care work is also evident in some of our most intimate and personal experiences. Seniors and people of all ages in hospice care find meaningful peace and strength in communion with an animal, as do those dealing with grief and trauma. Cynics may dismiss this as innate or fail to see this as work, but, again, the animals are asked and expected to be in particular places and positions, to behave in specific ways, and to subvert their feelings or desires in order to meet the needs of people; that takes and is work, and provides yet another example of animals' emotion work. Even experienced therapy dogs sleep deeply after being read to for half an hour because of the psychological stimulation of such a task.

Dollars and Sense?

Animals do not work for monetary pay in any conventional sense. Money itself is of no use to animals personally, yet domesticated animal workers may be "paid" through different means. An animal's labor may be compensated through direct food rewards for successful completion of a task, affection or praise, and/or the ongoing provision of food, water, and shelter. Critics may argue that this is not about an exchange, but rather a utilitarian need to keep the animal alive and working. Well-intentioned people are likely motivated by some notion of reciprocity, however, and Jocelyne Porcher (2014, 2016) argues that a sense of responsibility is at the heart of most human-animal labor relationships.

The idea of compensating animals for their work can also be approached in indirect ways. In many instances, the humans in possession of the animals who are doing work may be paid, and, in general, some of that money is spent on providing for the animals' needs and wants (the degree to which and its quality vary greatly). In certain jurisdictions, upon retirement, working dogs, especially police dogs, are afforded what could be called a sort of pension. For example, in the town of Nottinghamshire in England, people who provide homes for retired police dogs (often the officer and her/his family, but not always) are given £500 per year for three years by the police force to help cover medical costs (BBC 2013). In some instances, people who use service dogs, especially if the human is from the military and using a dog for posttraumatic stress assistance, are also provided with some public funds to cover the dogs' individual needs.

What is clear and undeniable is that domesticated animals' contributions to both formal economic processes and the quality of life of others are significant. Feminist political economists point out that unpaid social reproductive labor creates a great deal of "value" and provides a massive subsidy to families, economies, societies, and corporate interests, by continuously ensuring healthy, effective workers. Similarly, animals' work, both formal and informal (e.g., in homes), provides substantial benefits to people, economies, societies, and corporate interests everywhere. Thus we can enlist feminist insights not only to understand animals' social reproductive labor in the wild and in human-created communities, but also to think more deeply about pay, value, and how animals' work contributes to societies and economies. Moreover, political economists highlight how capitalism depends on the extraction of surplus value from people's labor; critical animal scholars like David Nibert (2013) argue that the exploitation of nature and animals is the foundational engine of capitalist exploitation. National and transnational for-profit industries use and "render" billions of animals bodies not only symbolically, but literally in painful ways (Shukin 2009). Whether and how we should think about these processes as involving animals' work is complex and warrants careful thought. Animal rights theorists and researchers highlight a number of industries seen as exploiting animals. They argue that using animals for entertainment, food, medical or cosmetics research/ testing, and sport is exploitive. Consequently, industries and events like horse racing, rodeos, zoos, circuses, and other media products for which animals are used (e.g., movies, greeting cards) are criticized, and often advocates call for their abolition. Animals are put to work in a number of these industries. Animal performances in circuses or for television are work. In sporting contexts such as racing, polo, cutting, jumping, and dressage, horses work.

The term exploitation is used in different ways, thus its meanings warrant elucidation. Exploitation is often associated with Marxist theory, and, succinctly, means use for profit, particularly use of someone's labor. It is used to describe both social relations and structural processes. For example, owners and employers exploit groups of workers, and, at the same time, the entire working class is exploited within capitalism because their labor serves and enriches capitalists. Thus, technically, every animal who works in a profit-making context meets both of these definitions of exploited because they and their work are being used to enrich others and the overall capitalist system. Some readers will object to all uses of animals' labor on these or other moral or ethical grounds. Some will also suggest the use of the term oppression to point to both individual and larger processes of subjugation, subservience, and hardship (Nibert 2013, 2014).

The concept of exploitation has also been taken up in different or expanded ways, politically and colloquially. It may not be used to mean use for profit, but rather refer to cruelty, coercion, and/or the process of taking advantage of someone in a way that is deemed socially unacceptable. Thus, someone who pursues a career they enjoy, even if working for pay and/or within a larger capitalist economy, may not see themselves as exploited. Overall, ideas of kindness and/or choice are interwoven with most socially accepted views of what makes "use" appropriate. Both kindness and choice are complex issues when thinking about animals' work. However, choice is also complicated when thinking about many people's labor (Lehmann 2007; McGrath and Strauss 2015). Most people are not truly "free" to pursue rewarding labor of their choice, not only because of their need for wages, but also because of structural constraints on their work-lives. These include discrimination, a lack of cultural, social, or financial capital, systemic unemployment, underemployment (that is, working for less than you would like—either the position or the number of hours), the expansion of precarious jobs, forced migration, and so on.

Choice is also linked to the concept of agency. Agency, understood concisely, means the ability to think, act, and make choices and/or change. Agency can be individual and/or collective. Dogs who live with people may choose whether or not to provide care and protection work, for example. As noted in the introduction, dogs are members of a species that likely chose to link themselves to people in the first place, thereby also actively participating in their own domestication. However, even dogs do not choose all their occupations or the subtypes of work therein. They may enjoy "working" in an active, formal sense, but once in an occupation, they generally must walk ahead of a military regiment into potentially dangerous territory, help apprehend suspects, be used against protesters or striking workers, or complete any number of other specific tasks, regardless of what they feel about these situations. There are some possible exceptions to this pattern that I will note below.

Most species did not choose to become domesticated or to work for people. Some individual animals put to work have been "tamed" (such as tigers in circuses) but they are members of species that have not been widely domesticated. Moreover, even these individuals are continuously subjected to measures of control to promote the desired temperament and behaviour (with varying degrees of effectiveness). Individual animals in such situations will continue to behave in ways that cannot aptly be described as "tame," but rather as expressions of natural behaviour and/or of agency. To get animals to work, force and punishment may be used, or animals may be cooperative and willing. Once in a situation of work, animals exercise their agency in a range of ways (see, e.g., Coulter 2014a; Hribal 2010; Nance 2013a; Onion 2009; Scott 2009; Shaw 2013; Thompson and Birke 2014; Warkentin 2009). They may be collaborative, form strong bonds and partnerships, and embrace the breadth of the tasks involved in their jobs. Some proactively act to protect, shape, or prevent, by drawing on their training and/or by enlisting their own intelligence. This proactive work may mean the animal disregards or disobeys human instruction to do what is actually more needed or appropriate given the situation. Moreover, some animals defy and/or obstruct, not to do a better job, but to resist their situation, particular tasks, and/or those mandating their work. How human workers react to animals' expressions of agency, especially those involving resistance, vary. People may respect animals' perspectives or they may force the animal to work, regardless. Animals' defiance can lead to pain or death.

Telling examples of these complexities, even for dogs, come from the experiences of the canines tasked with military work. Animals of all types and sizes, from pigeons to camels and elephants, and especially horses and donkeys, have a long and complex history of work for armed forces and in war zones (Hediger 2012; Nocella, Salter, and Bentley 2013). They have been tested upon, provided comfort and companionship, and been tasked with providing transportation and hauling labor, communication services, and detection work. Animals have also been weaponized, both by formal armies and by individuals who are resisting occupiers and/or foreign forces. Hundreds of millions of animals have been killed without fanfare, while others have received commendation for their contributions, while living or posthumously.

The US military has about 2,500 military working dogs and gives canine officers formal titles, often one rank above their handler's. Deep bonds between human soldiers and the dogs are often formed, and many continue in retirement after their formal working lives have ended. The dogs are responsible for various kinds of work, especially detection of explosives. They have safely found many bombs and saved many people's lives. Dogs have also been killed. Military working dogs are exposed to serious physical risks, resulting directly from the wars, as well as from the environment and exposure to different germs and viruses. Military work also takes a psychological and emotional toll on the dogs. Following tours in Iraq and/or Afghanistan, like their human counterparts, some military working dogs (an estimated 5 percent) suffer from posttraumatic stress. While deployed, certain dogs become highly stressed, distressed, and noncooperative, After reporting from Afghanistan, Michael

Phillips shared this powerful story: "Zoom, another Lab, refused to associate with the Marines after seeing one serviceman shoot a feral Afghan dog. Only after weeks of retraining, hours of playing with a reindeer squeaky toy and a gusher of good-boy praise was Zoom willing to go back to work" (Phillips 2010; see also Dao 2011, Paterniti 2014).

Some argue that the use of an animal's abilities, even if far away from war zones in safer, not-for-profit, charitable, and/or public sector spaces, is still exploitation or could lead to exploitive relations. Service dogs assigned to individuals have different lives than those who are taken into facilities for a few hours at a time, or even than those who live within care homes but who have a fair bit of control over their movement and work, for example. Scholars debate whether this kind of continuous service/care labor is an acceptable use of animals, or whether it violates their physical integrity, social lives, freedom, among other factors (Weisberg n.d.; Zamir 2006). Animals assigned to individuals full-time are tasked with year-round work, but generally given short breaks on a daily basis to act in ways of their choosing and relinquish their service responsibility. The dogs are given specific commands and/or held in particular ways (e.g., on a leash in contrast to a harness) to identify this "break" time. Yet there are few protections in place to monitor the lives and conditions of service dogs once they are in someone's private home, and the measures that do exist certainly cannot monitor treatment 24 hours a day and 365 days a year. Most people who employ service dogs do not harm the animals. Yet cruel treatment is possible and occasionally evident, even in public. People, and especially women, in personal support work, domestic work, and nursing are also subject to such risks, particularly if they live with their employers and/or are migrant workers. The more animals are put to work, the more there is a need to reflect on what the ethical and practical implications of labor are for the animals, what measures could or should be taken to protect and benefit them both during and after their formal work-lives, and whether all, some, or only a few animals should even be working for people. What is socially acceptable is contested, fluid, and context-specific. It is also shaped by what people know and do not know about animals' lives.

Toward Nuanced Understanding

Jocelyne Porcher has argued for scholarship focused on understanding what work means to and for animals (2014, 2016). Individual animals will have different feelings about their work and lives based on their personalities, health and well-being, breeds, the people for whom they work, and their perceptions and agency, among other factors (see, e.g., Lefebvre, Diederich, Delcourt, and Giffroy 2007). Overall, these dynamics reveal a complicated set of issues to consider when thinking about animals' work and the concept of exploitation, at individual, interpersonal, social, and structural levels. Accordingly, further recognition of nuance and of context-specific factors is prudent. An industry, particularly one that is for-profit, may rightly be called exploitive, in both the technical meaning of the word, and/or in the more subjective sense of being cruel. The purpose and structure of industries and workplaces encourage certain practices and ideas, and discourage others. However, this does not mean that every interaction is unpleasant, or that every (or even most) human workers within the industry seek to do harm or to exploit animals. As outlined in chapter 1, many people who work with animals across sectors do so because they care about animals. They may also engage in a range of daily behaviours at work that promote kindness and compassion, and contest patterns of exploitation, precisely because they "love" animals. The word love is a very political and significant metaphor and mobilizing force in animal communities and workplaces with many meanings and interpretations.

Therefore, the term exploitation may speak to the structural level, but obfuscate or even deny individual, interpersonal, and social dynamics that are more complicated. Similarly, as noted in chapter 1, individual feelings and even small actions can and do make daily life more pleasant for animals, but they cannot fundamentally change an exploitive industry or necessarily prevent a specific practice. However, the difference they can make for individual animals and their experiences should not be overlooked, a point Diana Stuart, Rebecca Schewe, and Ryan Gunderson (2013) also make. There is potential for daily, individual acts to be inter-woven with larger, more substantive attempts to make change as well. This important dimension will be taken up in more detail in subsequent chapters.

Accordingly, to combat homogenization, and capture dynamics that are more complex than a rigid exploitation-freedom dichot-omy, I posit a continuum of suffering and enjoyment for concep-tualizing animals' work for humans (figure 2.2). This recognizes the importance of context and of being nuanced. I have chosen the words suffering and enjoyment for two main, related reasons. First, these concepts reflect processes, understood temporally, and as lived experiences. Pain could be continuous or acute and fleeting, whereas suffering evokes awareness of embodiment, a longer-term dynamic, and the possibility of physical, psycho-logical, emotional, and intergenerational harm. Similarly, enjoy-ment can be both an immediate and an ongoing process. This fact intersects with the second reason for these terms, and that is animals' sentience. Both suffering and enjoyment emphasize ani-mals' abilities to think and feel. Suffering highlights the damage, the unpleasant, the horrific. Enjoyment reflects the fact that work can be a positive experience. It recognizes the breadth and diver-sity of animals' real working lives and experiences, and the fact that animals are whole and complex beings who can and do expe-rience joy, even at work. Animals are not fixed, biological entities who have remained unchanged over the last centuries and mil-lennia, and who can only experience enjoyment if returned to a previous, pre-human-contact state. Individual animals and whole species have changed and continue to change, physically, intel-lectually, emotionally, and socially, for many reasons, including due to their interspecies interactions and lives in human-created social communities. Thus, the poles of suffering and enjoyment, and, especially, the continuum in between, are crucial concepts for better understanding animals' work.

Figure 2.2   Continuum of Suffering and Enjoyment

Where animals' work fits on the continuum will be affected, in particular, by (a) the occupation (b) the work required (c) the employers or coworkers (d) the species and (e) the individual animals' own personalities, preferences, feelings, and agency. Some occupations mandate confinement, displeasure, degradation, harm, and/or fear. Others involve a greater degree of mobility, autonomy, flexibility, and enjoyment. Animals are asked (or required) to do work that is quite natural (e.g., cats being housed in a barn, library, or museum to hunt rodents), that builds directly on their innate abilities (e.g., detection and herding work for dogs), or that has nothing to do with who they are and what they do normally (chimpanzees on roller skates). The body, dirty, communication, care, and emotional work requirements also shape animals' experiences of work. Similarly, how the animals are treated by the people around them figures, and human workers make animals' work days and lives better or worse depending on their actions. Certain species are clearly more keen to work or willing to work with/for people, with dogs fitting the former, and equids the latter. However, individual members of larger species can vary greatly, and specific animals have different ideas and feelings about work, overall, as well as specific tasks. Certain dogs are enthusiastic about interactive, therapeutic work like visiting long-term care homes. Others are less social and not keen to interact with people beyond those they know well; care and protection work for their families is fine, but nothing more. Still others are very energetic and potentially even less suitable for or interested in a quiet home life. Conservation Canines, an organization based out of the Center for Conservation Biology at the University of Washington, for example, is one of a few organizations that seeks out shelter dogs with high energy drives who may not be appealing to families, thus less likely to be adopted. The people of Conservation Canines work with such dogs to train them to detect signs (like scat) of endangered species, allowing human researchers to gain knowledge about their behaviour, health, population size, and range. These dogs' abilities and energies are thus channelled in the pursuit of conservation efforts. Dogs are working on conservation projects and on anti-poaching brigades around the world, including in Latin America and Africa.

Animals' feelings can also vary depending on the moment in time, the situation, and other factors. A normally positive animal may have days or even longer periods of discomfort or disinterest stemming from their physical or emotional state, the tenure of their work, or a negative interaction with a person, another animal, or something in the environment. An animal usually reluctant or timid may also be empowered by a new human worker who shows concern and compassion. Enlisting the continuum allows for recognition of context-specific factors, as well as fluidity and change for the same animals.

Horse racing clearly illustrates this continuum. It is a sporting industry that relies on a high number of horses and working-class people, and, objectively, the intense use of both; people and horses are worked hard, and the work is physically risky for both. Racing is considered cruel to horses by many animal advocates. The people involved in racing as workers have uneven, varied, and sometimes contradictory ideas about the industry and how both people and animals are and should be treated. Human workers can express a love for the sport, horses, and culture, frustration, and anger with practices or treatment, and/or suspicion toward outsiders who critique these dimensions (Case 1991; Cassidy 2007; Scanlan 2006).

Horse racing is a larger umbrella under which a number of industries and types of competition fit. Horses are tasked with different occupations within racing. There are breeding horses, horses in training, and active racers (this role lasts no more than a few years in most cases). Horses generally live for 20 or 30 years, and certain racers go onto second careers as pleasure horses, show jumping horses, dressage horses, or breeding horses, while others are euthanized or sent to slaughter right after they are no longer used for racing. This can be as early as when the horse is four or five years old, or even sooner if a serious injury is either deemed catastrophic or likely to require many months of costly treatment and nonracing time.

Among the active racers, there are different breeds, including Thoroughbreds who do flat racing or steeplechase (jumps racing) with jockeys on their backs, Quarter Horses who do shorter flat racing also with riders, and Standardbreds who do harness racing, pulling drivers in sulkies. The people involved with each type vary, including in their class locations and their ideas about horses. The rates of injury and death are generally higher for Thoroughbred horses, especially those in steeplechase racing. This is influenced by the physiology of the animals (very fine-boned), the young age at which training and racing starts (often two years of age), and the labor required (leaping over jumps on a race track at high speeds with about a dozen other horse-rider pairs). Dozens of actions taken (or not taken) by human trainers, grooms, and jockeys affect horses' health, bodies, and well-being, on and off the race track (see, e.g., Beisser et al. 2011; Case 1991; Cassidy 2007; Williams, Smith, and DaMata 2014).

While the animals are all of the same species, how individual horses feel about racing varies substantially. Some horses display physical and behavioural indicators of enjoyment and are singled out for loving to run, while many others exhibit signs of discomfort, pain, anxiety, anger, and so forth. Some horses simply refuse to exert themselves on the race track and what happens to these objectors varies greatly. Through my ethnographic research on show jumping cultures, I have witnessed stark differences among individual horses, as well. Certain horses thrive in situations of competition; others clearly dislike their jobs and lives; many fall somewhere in between and/or their feelings change over time. Horses communicate about their physical and emotional states to those who are interested in understanding. As explained in chapter 1, horses' ears, their body weight and condition, their body positioning, the look in their eyes, their degree of interactivity, and a number of other factors speak volumes.

Working for a Living or Worked to Death?

Tens of billions of animals reside on/in industrial livestock operations where they will be used to create products that are edible and/or wearable; these animals warrant careful consideration. Arguably, these animals are positioned within a perpetual cycle of disposability, the specifics, degree, and length of which vary. Some animals' bodies are used in their entirety to create various kinds of meat, leather, fur, food for pets, and other purposes. Other animals' whole bodies are deemed irrelevant to the process because of innate features like sex or due to health and injury. Some are killed mere hours after their births (male chicks deemed superfluous by humans in the egg industry because these birds cannot produce eggs), a few months later (calves used for veal), or after a few years (cows and pigs used for breeding and/or meat). Some cows are designated exclusively for beef and generally are kept alive for just under two years. Cows who produce milk are also slaughtered for meat after a few years when they are considered "spent" from continuous impregnation and milking.

The ability for these animals to engage in subsistence work and/ or social reproductive labor varies. For some, their food is provided, while others engage in self-feeding through grazing in set areas. For certain animals, the females raise their own offspring before they are sent to slaughter (such as on beef farms). Many other animal mothers are prevented from interacting with their infants for more than a few hours or days. For example, on most dairy farms in countries like Canada, male calves are promptly sold to veal farms or put into veal hutches or crates that may be visible to and in the hearing range of cows, but out of their physical reach (calves may or may not be able to touch and interact with other calves). Most female pigs in North America are kept in gestation crates for most or all of the time they are used as breeders (i.e., for a few years). Piglets can nurse and be near their mother for a few weeks or months before being sent to a "finishing barn" or a slaughterhouse. The mothers can lie down and stand, but they cannot move or turn around to nuzzle or lick their piglets. These are the dominant practices in contemporary industrial agriculture, although there are a small number of farms that follow different practices (and gestation crates are already illegal in certain places including the European Union, and are being phased out in others). The exceptions may or may not be organic farms; an organic designation can include aspects of animal treatment, or it may focus on factors that most clearly affect the humans consuming the animals such as what medications and hormones animals were or were not given.

The ability of animals in most industrialized contexts to provide their offspring with even basic biological needs is contingent and constrained, or eliminated. Their ability to engage in species-specific behaviour, intergenerational educational interactions, caring work, and other forms of social behaviour is very or entirely limited. Barbara Noske (1989, 1997) enlists Marx's concept of alienated labor as a lens for understanding these animals. The concept of alienated or estranged labor refers to how workers are materially and spiritually separated or disconnected from the products, results, and/or rewards of their labor, as well as from their desires, or what Marx calls their species-being. Noske (1989, 18–20) argues that animals in industrial agriculture are alienated from: (a) the product (their own offspring or parts of their body) (b) productive activity (not being able to turn around, for example) (c) fellow animals and their social nature (d) surrounding nature and (e) their species life. Referring to animals in these situations in this way recognizes that they do have work assigned to them such as creating babies, producing milk, and laying eggs, but that these are not contexts of choice or much (if any) pleasure. Instead they are examples of alienated labor (Noske 1989, 1997; Stuart, Schewe, and Gunderson 2013). Advancing ecofeminist analysis, Carol J. Adams (2010), Karen Davis (1995), and Lori Gruen (1993) rightly point out that agriculture involves a highly feminized form of production, as female animals and their bodily processes are disproportionately used and manipulated (see also, Gillespie 2014). As noted, young male animals also often face very quick deaths.

Jocelyne Porcher and Tiphaine Schmitt (2012) propose another way to think about animals' work in farming contexts. Based on field research and direct observation of both farmers and animals, they argue that cows producing milk invest their intelligence, collaborate, cooperate (with each other, farmers, and/or the milking robot) adapt, and cheat, thus actively and subjectively engage in daily work and the larger process of milk production. This dimension adds another layer of understanding, one that focuses on a farm-specific labor process in which the cows are involved, even if their primary job is to produce milk for human consumption. Without question, the degree to which cows are able to engage in these kinds of active processes varies depending on the structure, size, and organization of the farm. Porcher and Schmitt's (2012) study focuses on a smaller, farmer-owned operation and is contextualized within the longer history of French farming culture. While the majority of farm animals today in countries like Canada are kept in industrial operations, a global perspective reveals greater diversity in structures and practices, and these have implications for animals' lives. Living with a peasant family that tends a few goats and even being a pig on larger farms in contexts where gestation crates have been banned translate into both different and similar experiences for animals. It is not as simple as smaller equals better, however, and the smallest-scale peasant farmers may cherish the working animals in their lives, actively abuse them, or cause discomfort or pain unintentionally due to a lack of knowledge and/or because they lack proper equipment and resources. An animal pulling a plough is both similar to and very different from an animal whose job is to lay eggs inside a tiny, cramped battery cage, stacked on top of, below, and beside thousands of other hens. Again, context and specifics matter. Overall, there are arguments to be made that animals engage in work on farms/in agriculture, but this is an area that warrants care and caution, and questions of choice, agency, exploitation, and the continuum of suffering and pleasure should be kept front of mind.

There are more animals on farms than anywhere else on land (excluding invertebrates), and these animals' daily lives and deaths are largely hidden from social view. Outside of agriculture-specific discussions, these animals are not widely considered, period. The conditions of their lives are, in most cases, different from individual animal athletes like race horses, who are indeed viewed as commodities, but simultaneously as individuals with personalities, moods, and abilities—a strong illustration of Wilkie's (2010) concept of sentient commodities examined in chapter 1. As Wilkie argues, farmers can see animals on a commodity-continuum, but this is particularly applicable to smaller or family-owned farms, and less the case in industrialized agricultural facilities that dominate the North American landscape. Every race horse is given a name and often a nickname; very few of the billions of animals raised and killed for food have their individuality or sentience recognized in any way. At the same time, an argument could be made that the exploitation of animals making or intended to be food is a more truthful process, in contrast to those who treat animals as named individuals while they are generating money for humans, only to then dispose of those beings once their use has ended (through sale, euthanasia, or slaughter). All told, there are commonalities and differences among working animals across a number of for-profit contexts, and Noske's animalization of the concept of alienated labor could fruitfully be applied to a number of animal work spaces where it is supported by evidence.

Animals in agricultural contexts are producers of a kind, and as alienated as they may be, and as disembodied and disconnected as they may be from the final products of their bodies (and, as technocratic and mechanistic as this sounds), they are still part of the "production process" itself. The animals required to produce babies, eggs, or milk are doing a kind of work. Their bodies are the materials of production, and they are also often alienated in the sense proposed by Noske. When an animal's sole "task" is to eat and fatten in order to physically become meat or fur, I am uncertain about whether to argue that they do work, other than perhaps some very constrained and limited subsistence labor to sustain themselves and their offspring.

Overall, in agriculture, animals' bodies play a central role, yet it is different in key ways from the body work examined earlier in this chapter. Given the breadth, complexity, and significance of ways animals' bodies are involved in the cross-section of their working lives, there is good reason to thoughtfully develop an animal-centric concept of body work, by building on but going beyond labor process theories and other humanist conceptualizations. Such intellectual labor should recognize the sexual divisions and implications, and grapple with whether animals whose bodies are the only product that will be consumed should be seen as working. Without question, the workplaces and industries within which most animals' lives are almost exclusively about suffering, extreme instrumentalism, and indignity in the pursuit of profit are complicated and they are troubling. Framing the animals therein as workers may or may not be conceptually, ethically, and politically useful. Some propose slavery as a more accurate term, while others caution against the widespread enlistment of this term for historical and sociological reasons, due to questions of appropriation, and/or as part of a broader call for decidedly posthumanist language rooted in animal-centric specifics. Drawing on postcolonial and antiracist feminist theories, Lauren Corman (2013) also argues against monolithic statements of victimhood that construct animals (or people) as entirely voiceless, even in situations of deep oppression. Of course, enslaved people work, and acknowledgment of coerced labor does not prevent the accompanying use of supplementary and corollary concepts to further describe, contextualize, or unpack such situations. We can recognize work done by animals of all kinds, and additional specifications and contextualization refine and reflect the differences.

These complex dynamics are clearly apparent in workplaces where animals are held and used for experiments. This domain has been studied from a number of angles as noted in chapter 1, and, interestingly, whether animals' perform work in this sector has been give a relatively noteworthy amount of thought, particularly by those interested in biopolitics and animal ethics. Donna J. Haraway (2008), for example, moves easily from considering dogs in "state jobs" as airport security providers, drug and bomb sniffers, and the like, to seeing dogs in laboratories as workers. In contrast, Zipporah Weisberg (2009, 37) rejects labels of workers and even alienated workers for those in labs, seeing these animals as "worked-on objects, *slaves* by any other name. To call them anything else is to gloss over the brutal reality of the total denial of their ability to act in any meaningful way." Most recently, Jonathan L. Clark (2014) applies the concept of clinical labor to both human "guinea pigs" and actual guinea pigs

I remain hesitant about suggesting that experimentation on animals is an example of work done by animals. Dogs sniffing for drugs in airports and those kept in a cages as test subjects, are all being used by people, but there are substantive and substantial material and experiential differences between these two situations, particularly if seeking to consider the dogs' perspectives and feelings. Animals' bodies react to what has been administered, injected, irradiated, implanted, removed, or opened, thus animals respond to the body work performed on them by humans. Animals also react to experimentation processes through their obedience, their (ultimately unsuccessful) avoidance, and/or through acts of defiance, and they struggle daily to survive and cope with the physical and psychological damage done to them. But they are not working.

It is worth noting that research on animals can mean and involve different things. Some animals being housed in research facilities may be asked to complete specifics tasks and these could more clearly be seen as work. For example, the fairness studies initiated by Sarah F. Brosnan and Frans de Waal (2003) and expanded since (see, e.g., de Waal and Suchak 2010) include giving the animals "a job," even if it is as simple as taking a rock and handing it back to the researcher. The animals are paid with food, and deliberately paid inequitably (with one animal getting visibly better payment such as a more desirable item) in order to assess their responses and collect data on animals' perceptions of fairness. The results are fascinating and many animals greatly object to unfair pay. Moreover, some who are provided with better pay have actually stopped doing the requested task, presumably to protest the fact that the other animal is receiving lower pay, a remarkable act.

In laboratory contexts where animals are not to complete tasks, but rather are to be the physical recipient of tests, I am more persuaded by Weisberg's (2009) emphasis on the animals as being acted upon. At the same time, I do not see "slaves" as an entirely accurate descriptor either. First, I am cognizant on antiracist scholars' and historians' insistence of avoidance of the term "slave" even for human-human situations of enslavement, and of the importance of using "enslaved people" instead to foreground that these are/were people, first and foremost, whose identities as such should not be discursively obfuscated or erased. Second, enslaved people were primarily enslaved for their abilities to work; they were also enslaved workers. Animals kept in laboratories are not kept for their ability to work; they are animals on whom experimentation and testing is conducted. They are also individuals who are physically, psychologically, emotionally, intimately, and sometimes fatally harmed by what happens to them. Thus, despite the difference in species, examples that seem more comparable are the cases of experimentation on people deemed subordinate due to their race, ethnicity, or developmental status. These situations are not identical but they are, arguably, more similar than other comparators. Admittedly, my thinking on these questions is ongoing and incomplete. Most importantly, even if these are not examples of work done by animals per se, ethical positions and actions can still be being taken. In these cases, our intellectual desires to understand and categorize labor are not the top priority. As Lynda Birke (2009, n.p.) puts it, "Animals may indeed be supremely indifferent to the names we give them: but they are not indifferent to the naming of oppression."

There remain many intellectual and political reasons for examining and thinking through the breadth, depth, and impact of work done by animals across contexts, in all its complexities and unevenness. Doing so counters totalizing generalizations and ill-informed statements of all kinds, and instead encourages the pursuit of empirical data, accuracy, and nuance, thereby providing fodder for more thorough knowledge and understanding. Such an intellectual commitment helps write animals into history and social and political thought, extending the call from those in labor studies and other critical fields to consider marginalized actors beyond its anthropocentric and thus incomplete focus.

There are, notably, also expanded possibilities that can become more visible and enactable when animals' contributions are acknowledged. Rosaleen Duffy and Lorraine Moore (2010, 761), for example, argue that elephant trekking has changed how the animals are both viewed and treated in the community studied because "they can now be valued as a source of labor." I will revisit to what degree animals' movement into and evolution within worlds of human-organized work helps and/or harms animals in the concluding chapter. Seeing animals' work could also have conceptual and more broadly political implications. As Jonathan L. Clark (2014, 157) puts it, "Refashioning the category of labor to include nonhuman animals helps challenge the paradigm of human exceptionalism that justifies so much violence against animals." Writes Donna J. Haraway (2008, 73), "My suspicion is that we might nurture responsibility with and for other animals better by plumbing the category of labor more than the category of rights…Taking animals seriously as workers…might help stem the killing machines."

Whether a labor politic is actually more useful or effective than other approaches and frameworks is hard to know, especially given the indifference or hostility of some people to the struggles and plights of their fellow human workers whose well-being and worth are generally prioritized above that of animals, and considering the current refusal on the part of so many labor advocates to expand their spheres of empathy beyond humans. Yet the politics of animals and work are both predictable and unexpected. For example, with his novel, *The Jungle*, which focuses on slaughterhouse work and workers, Upton Sinclair says he "wished to frighten the country through a picture of what its industrial masters were doing to their victims; entirely by chance I had stumbled upon another discovery—what they were doing to the meat supply of the civilized world. In other words, I aimed for the public's heart, and by accident I hit it in the stomach" (McChesney and Scott 2003, x). People have complex political ideas, and a lack of regard for the labor and lives of some people may not necessarily translate across species. It is also possible that more labor advocates' ideas about animals will change. Women, workers of color, LGBTQ workers, migrant workers, among others who did not fit the white, male, manual labor model of a standard worker were initially excluded from not only definitions of who was a worker, but many workers' organizations and webs of solidarity. Today, while certain researchers and workers' advocates continue to exclude even these kinds of human workers (through deliberate or neglectful means), doing so is widely seen as unacceptable and problematic, and it is at least officially condemned by the majority of workers' organizations and activists. All told, given the realities across the planet today, every strategy with potential warrants serious consideration, and I sincerely hope that thoughtful consideration of animals' work is not only conceptually but politically and practically useful.

The processes of identifying and understanding animals' work are crucial initial steps, and this chapter has shared ideas for how to do both. I have proposed the employment of existing concepts from labor process theory and feminist political economy, but also suggested promising avenues for further developing animal-centric analyses, including through the fostering of an expanded and more-than-humanist concept of body work. Multispecies work involves affective labor, as multispecies bodies, emotions, and senses are all engaged and affected (see, e.g., Parreñas 2012). Thus, along with body work and ideas of intimate labor (see, e.g., Boris and Parreñas 2010), the affective lens may offer some fruitful intellectual potential. In any event, the specifics of dirty, communication, care, emotional, and other forms of work are valuable for elucidating the particulars of animals' contributions, and really unpacking the labors involved.

In sum, I have posited three main organizational categories that reflect the full breadth of work done by animals—for themselves, voluntarily for humans, and as mandated by people—and given this final category the most intellectual attention. I have argued for context-specific analyses that consider a continuum of suffering and enjoyment, as well as ideas of exploitation and alienation. Crucially, the process of *recognizing* animals' labor is the next essential challenge, one that also involves work.

3

The Work Done With/For Animals: Political Labor and the Work of Advocacy

The third subcategory of animal work is also done with/for animals, but it involves a more expressly political kind of labor. For some human workers, their days involve direct interactions with animals and the provisioning of care work, yet this is only one part or manifestation of a broader commitment to working for and with animals; they also engage in advocacy work. For others, they may rarely, if ever, directly interact with animals, yet their work-lives revolve around advocating for and/or protecting animals. Joan Tronto (1993, 172) posits that care can become "a tool for critical political analysis when we use this concept to reveal relationships of power." In that spirit, political labor and the work of advocacy is the focus of this chapter.

Political action is a core consideration in much labor studies scholarship and for good reason. Writers often examine how and why workers seek to reject, resist, or change the conditions of their employment, their employers, and/or the ideas, practices, and systems that create both daily and sustained, structured inequities and unfairness. A commitment to inquiry that seeks to understand processes of resistance and to generate knowledge that can propel projects of social change builds on Karl Marx's (1978, 145) argument that "philosophers have only *interpreted* the world, in various ways; the point, however, is to change it." Given the conditions of many animals' lives—and because labor studies analyses of political action have, so far, been very anthropocentric—there is even greater need for a multispecies approach to both understanding and fostering political action.

At the same time, this third type of animal work also stems from the data. In examining the breadth of work done with/for animals, the amount of time and energy that many human workers allocate to decidedly political work is noteworthy. A substantial amount of labor and entire occupations involve making life better for animals. The efforts can be focused on a specific group in a particular place, a cross-section of species in a shared geographic region, an entire species, or a purposeful commitment to "all animals." Human workers actively envision and pursue strategies that seek to expose, question, challenge, and/or change the conditions of animals' lives, and the causes of suffering. The types of political work pursued vary greatly and thus warrant further examination and unpacking, a task I will begin in this chapter.

Politics, regardless of its focus and organizational base, is work (Coulter 2011; 2014b; Coulter and Schumann 2012; Franzway and Fonow 2011). Political work includes explicitly partisan or government-focused initiatives, but is also much broader. Political labor can be done in governments of all levels, formal parties, nongovernmental organizations of all kinds, community groups, labor unions, smaller networks, informal clusters, collectives, coalitions, councils, and/or individually. It may be paid or unpaid, with the latter involving volunteers and/or activists. Every political project relies on people to develop and contest ideas, organize, resist, negotiate, challenge, implement, and/or assess, as well as carry out mundane but essential tasks. In other words, political projects only exist because people do the necessary work. Different kinds of work are required, depending on the specific political project or organization.

The political work examined in this chapter is approached as multispecies and interspecies. The human work done in/for animal-centric efforts is first considered. I then turn to human-centric advocacy strategies that are used in multispecies workplaces and consider whether and how animals are included in these advocacy efforts. Specifically, I highlight the two most widespread and pertinent advocacy vehicles used when thinking about animal work politics: nongovernmental or advocacy organizations, and labor unions. I conclude by introducing the public sector as a current and future space of advocacy work.

When examining efforts to make change, the concept and practice of agency is again important. As concisely outlined in chapter 2, agency refers to the capacity to think, make choices, act, and make change. Agency and political work are complementary and these concepts foreground linked but also distinct facets of work done with/for animals. By enlisting the concept of agency, I am building on recent scholarship across disciplines that recognizes animals' subjectivities and that sees animals as social actors who shape their own days and lives, and those of others (see, e.g., Corman 2012; Coulter 2014a; Cronin 2011; DeMello 2013; Hribal 2007; McFarland and Hediger 2009; McHugh 2011; Smuts 1999). Notably, Sue Savage-Rumbaugh coauthored an article on people's treatment of bonobos with Kanzi, Panbanisha, and Nyota (2007), apes who could communicate through a computerized keyboard, in order to share the apes' views on the issue. This is why I have not proposed the examination of the advocacy work done *for* animals, but rather the work done *with/for* animals. Such a framing positions animals as subjects and agents, to differing degrees, in questions and projects of political change. Animal advocates continuously claim that they are speaking "for animals" and that animals are voiceless. Yet animals do have voices and other modes of expression; people must seek to understand what animals are saying or trying to communicate. Of course, many animals are literally prevented from expressing themselves through highly repressive and oppressive conditions. At the same time, as Susan Nance (2013a) points out, animals can and do exercise agency through different means, but they cannot comprehend the larger systems of power and stratification within which they are enmeshed (although similar could be said of some people). Animals can, however, understand and feel the effects of these structures on their own bodies and on fellow animals, especially those experiencing terror.

Yet the fact remains that our political and legal structures are designed for human participation. Animals cannot vote, lobby, organize demonstrations at legislatures or corporate headquarters, or create campaigns. Newborn rhesus macaques taken from their mothers for maternal deprivation experiments cannot sue the

University of Wisconsin-Madison, but the Animal Legal Defense Fund (2014) can, for example. Thus, human political work is the primary focus of this chapter, and it is essential. However, I also support arguments for expanded conceptualizations of agency and politics that recognize how animals express themselves, engage, and act. Moreover, animals are also involved, in different ways, in human-driven political work. Most significantly, there are different possibilities that can emerge from thinking through how animals are interwoven with political ideas and projects, and I will expand on these issues in the conclusion.

The concept of praxis is also relevant to this chapter. Praxis means action that is informed by theory, by ideas. Although it refers specifically to action, action and thought are understood to be connected. Specific steps are taken for intellectual reasons, yet the process of taking action also generates new or deeper insight. Thus, theories are, ideally, enriched and updated. In this way, the use of the term praxis encourages an ongoing thought-action-thought relational dynamic. Of course, while someone may hold a particular set of views or vision, this does not guarantee that any specific plan of action or political strategy must necessarily be pursued; there is not a deterministic relationships between theory and action. Those with a great deal of shared understanding or a comparable set of political beliefs may pursue and promote identical, similar, or very distinct forms of praxis. Social actors negotiate the relationships between their ideas and their political work in context, and are shaped by a range of factors including their strategic assessments, positions, job descriptions, workloads, job security, time and schedules, life stages, personal commitments and responsibilities, status, gender, race, ethnicity, class, citizenship statuses, confidence, personalities, allies, among others (Franzway and Fonow 2011; Gaarder 2011b). Nevertheless, ideas and analyses encourage people to see particular kinds of actions as advisable and worthy, and other practices or approaches as less desirable.

Accordingly, praxis complements the concepts of agency, advocacy, and political labor. All of these concepts reflect the process of trying to make change and advocate, but emphasize or highlight slightly different facets of the process. In order to examine political work involving animals, I will first briefly outline some of the primary ideas and theories that are informing forms of animal-centric advocacy.

Approaching Animal Advocacy

Much of the political work being done for/with animals is informed by a belief that animals should have rights. As Sue Donaldson and Will Kymlicka (2011) point out, however, there can be connections or noteworthy differences between the theories of animal rights generated by scholars, and the ideas of frontline activists and advocates. Moreover, frontline animal advocates, wherever they are working, themselves have a range of different views. There is some overlap but there are also stark differences; proponents of animal rights are not a homogeneous group. Some writers propose a dichotomy between animal rights and animal welfare ideas, but this is not sufficient either. Animal welfare proponents always believe animals have and should have rights; the debate is about which rights—and for which groups of animals. At the same time, there are people who work for the well-being or protection of certain animals and/or on specific issues, but who explicitly deny a connection to or association with animal rights activists and movements (see, e.g., Greenebaum 2009).

The roots of contemporary ideas and forms of praxis are diverse and context-specific. Different perspectives on people and animals, their roles, and their entitlements have been constructed, articulated, and put into action in many ways cross-culturally, and even a synthesis of the breadth of these data is beyond the scope of this chapter (see, e.g., Campbell 2005; Gaynor 2007; Hribal 2007; Hurn 2012; Kemmerer and Nocella 2011; Kalof 2007; Knight 2005; Linzey 2009; Ritvo 1987). What is especially noteworthy is that there is a long and rich history of individuals—from farmers to poets to philosophers—and whole communities that espouse some kind of connectedness with, respect for, and/or reverence for animals. These views have translated into individual or larger social actions that improve the lives of animals, as well, such as not killing them for food (see, e.g., Gregory 2007; Stuart 2006; Walters and Portmess 1999). Perceptions and practices have endured and/or have been changed over the most recent millennia and centuries.

The main kinds of decidedly political advocacy seen today have organizational roots in the mid- to late-nineteenth century, as do early examples of wildlife conservation work (Ingram 2013). Animal advocacy emerged in a context of growing humanitarian, progressive, and reform movements. Authors such as Jack London played a role, and Anna Sewell's 1877 novel *Black Beauty*, narrated by a working horse, inspired and propelled great empathy in readers. Although not as famous as *Black Beauty*, Margaret Marshall Saunders' novel *Beautiful Joe*, which was written from the point of view of an abused dog, also fostered compassion for animals. The Beautiful Joe Heritage Society continues to this day and there is a named memorial park in Meaford, Ontario. Many of the early Societies for the Prevention of Cruelty to Animals (SPCAs) and Humane Societies, including in England, the United States, and Canada, were motivated by a desire to curb cruelty to horses visible everywhere, pulling people and goods through the streets (Beers 2006; Ingram 2014; Kean 1998; Preece and Chamberlain 1993). The groups' activities were diverse and they confronted the treatment of animals on farms and in circuses, hunting, strays, animal products in fashion, among other issues. Humane education campaigns were created to proactively shift cultural ideas about animals, particularly among children and youth. Organizations and campaigns focused on ending vivisection were also very active and engaged (Cronin 2014; Ferguson 1998). "Lizzy" Lind af Hageby, an affluent Swedish feminist and anti-vivisectionist, targeted the male-dominant medical establishment, for example. A noteworthy campaign (and then a riot) against vivisection and in defense of an "old brown dog" who was subjected to vivisection for two months at the University College medical school in London, England, brought together early feminists and workers, groups that did not always collaborate (Lansbury 1985).

The ideological, gender, and class politics of early animal advocacy efforts were interesting and uneven, overall. Women and men organized separately and together, with women often doing most of the daily work, but men maintaining the high-status leadership positions (a pattern that often endures even today), although a number of women founded and led organizations including Caroline Earle White, Frances Power Cobbe,

Emily Appleton, and Dorothy Brooke (Beers 2006; Kean 1998). As is true today, women did the majority of the early animal advocacy organizing, particularly those from the middle and upper class, although George T. Angell, a cofounder of organized animal protection in New England came from a poor family, and working-class people advocated for animals in a range of ways in their daily lives, if not always through formal advocacy organizations (Gaynor 2007). Many emerging middle-class reformers, including J. J. Kelso in Canada, sought to protect both poor children and animals (Chen 2005; Rutman and Jones 1981). Notably, some early advocates for animals saw connections between the abuse of animals, women, and, sometimes, human workers, including enslaved people, illustrating an intersectional analyses of the causes and effects of oppression. It was not uncommon for advocacy efforts to frame animals as one of a number of oppressed groups, the others of which were human.

Historians differ in their interpretations of the classed dimensions of early animal advocacy work. Diane L. Beers (2006, 9) argues that "like many reform movements, this cause attracted middle- and upper-class men and women not so much because they had misguided obsessions or an overwhelming desire to socially control poor people but rather because they had the time and disposable income to support the diverse reforms they believed would uplift all humanity and protect nonhuman species." Indeed, early humane societies charged membership dues in order to be able to sustain their work, and given the breadth and depth of poverty at the time, most people would simply not have been able to afford such fees. Clay McShane and Joel Tarr (2007) point out that some members of the capitalist class, like Henry Bergh, founder of the American Society for the Prevention of Cruelty to Animals (ASPCA), were criticized for condoning or tolerating a number of cruel practices when done by members of their own class (such as tail docking), while emphasizing working class abuses of cart horses. McShane and Tarr (2007) and Ann Norton Greene (2008) also note that individual workers were regularly fined, while the employers who mandated increasingly demanding schedules and timeframes which pushed drivers to push the horses, were not held accountable.

At the same time, Diane L. Beers (2006) highlights a number of campaigns that pitted animal advocates from the upper class, including Bergh, directly against their socioeconomic peers' business interests, leisure pursuits, and fashion choices. Darcy Ingram (2014) notes that in Canada's interspecies history, many members of the capitalist class engaged with animals in different ways, including through sport, hunting, ranching, and breeding. These men (and they were predominantly men) were interested in the welfare of certain animals, Ingram argues, yet were simultaneously implicated in and often promoted the exploitation of other animals (a pattern not uncommon today). Around the same period and even earlier, some prominent socialists like George Bernard Shaw, Albert Einstein, and Henry Salt, linked exploitation of humans and animals within capitalism. George Orwell (quoted in Crick 1992, 451) wrote *Animal Farm* because he recognized that "men exploit animals in much the same way as the rich exploit the proletariat." Undoubtedly, the ideological and classed dynamics of animal work politics have never been simple.

The dynamic movements that were commonplace and highly engaged in the late nineteenth and early twentieth century became somewhat less visible, but also changed as decades passed. There were always committed groups of interspecies advocates, and more SPCAs were created, but their political work was not as publicly apparent, and often focused primarily on companion animals and stray dogs (Beers 2006; Finsen and Finsen 1994; Jasper and Nelkin 1992; Niven 1967). In the 1970s, there was a reinvigorated discussion of animal well-being (broadly conceived), inspired, in part, by Peter Singer's (1975) book *Animal Liberation*, and some strands of environmentalist and deep ecology thought. Animal ecofeminists also sought to identify gendered entanglements in harm and well-being, although their efforts were less recognized at the time, and are generally given less attention in historical writing (Adams and Gruen 2014b; Gaard 2011). There has been a growing interest from both scholarly circles and activist communities in animals' well-being and in how people ought to think about it since, including fierce debates about both theory and praxis. Without question, there is a much more textured and fascinating history than I have captured in this brief overview, and the histories referenced provide far greater insight into these processes and their contexts (see also, Montgomery 2000; Silverstein 1996; Singer 1998; Stallwood 2014). My goal here is to briefly situate contemporary efforts in longer historical and intellectual contexts.

In recent years, different strands of animal rights thought have been developed, and a number of prominent debates have occurred among proponents of particular approaches and analyses. As noted, advocates have different views about what rights are to be afforded, to which groups of animals, and for what reasons, and diverse perspectives about what forms of praxis ought to be pursued. Scholars of animal rights unpack and debate each other's analyses, assumptions, and proposals, as part of pursuing healthy intellectual debate. Akin to anthropocentric critical scholars and social actors, the animal rights literature is also infused with some fiercely antagonistic dynamics. A reader new to the literature—or to animal advocacy—may wonder if certain proponents of animal rights spend more time criticizing each other than they do targeting the interests perpetuating the harm against animals or working to broaden support for animals.

Nevertheless, it is helpful to frame the conceptual approaches themselves as something of a continuum. Animal rights can be thought of as an umbrella under which different approaches can fall. Succinctly, on one side is an animal welfare or a welfarist vision. This approach rests on the belief that some human use of animals is acceptable or inevitable, but that the goal should be to eliminate or restrict cruelty and suffering by providing animals preventative and positive rights (i.e., the right or freedom to, as well as protections from). On the other side, an abolitionist approach to animal rights is underscored by the position that animals require only one right: the right not to be human property (owned) (see, e.g., Francione 2008). In other words, this approach is driven by the idea of eliminating all human use of animals. Some proponents of this approach can also be called extinctionists, as they argue that the existing domesticated animals should be allowed to live out their lives, but that no further breeding take place. Such a position is sometimes called animal liberation—the idea of liberating animals from humans. Yet the term animal liberation has also been invoked by advocates like Peter Singer who enlists a utilitarian, philosophical, and applied ethical approach that argues for the pursuit of the greatest good for the largest number, and emphasizes animals' capacity to suffer as sufficient grounds to grant them rights. Consequently, there are debates about and different uses of the term animal liberation and this can cause confusion.

In between or in addition to these two spheres are various visions of animal rights, including the right to personhood, legal protections, and/or citizenship. Scholars draw from and have developed a number of linked and distinct conceptual visions for improving animals' lives including capability theory, deep ecology, feminist care theory, and ecofeminism (see, e.g., Cavalieri and Singer 1994; Donaldson and Kymlicka 2011; Donovan and Adams 2007; Nussbaum 2006; Regan 1987; Sunstein and Nussbaum 2004). The intellectual landscape of animal rights is vast and complex, and thoroughly mapping it is another ambitious task well beyond the scope of this chapter. The most salient insights to induce from this brief overview are that the term "animal rights" in fact refers to a heterogeneous and often divided set of ideas and social actors. There are diverse political ideas about animals, and these intersect with different forms of political work.

There are clear examples of particular actions that stem from specific ideas. For example, the "five freedoms," an established set of guidelines used across countries, originate in a British report written by the Farm Animal Welfare Committee, which sought to ensure basic welfare provisions for animals on farms, and, arguably, in other contexts:

1. Freedom from hunger or thirst by ready access to fresh water and a diet to maintain full health and vigour.
2. Freedom from discomfort by providing an appropriate environment including shelter and a comfortable resting area.
3. Freedom from pain, injury, or disease by prevention or rapid diagnosis and treatment.
4. Freedom to express (most) normal behaviour by providing sufficient space, proper facilities, and company of the animal's own kind.
5. Freedom from fear and distress by ensuring conditions and treatment which avoid mental suffering. (1979, 1)

These rights clearly delineate a welfare approach by accepting that animals are going to be kept in agricultural contexts and used for food production, but that they are nevertheless entitled to a decent quality of life. The five freedoms illustrate what may be called a "use but not abuse" approach to animals' well-being characteristic of welfarism. In contrast, an organization like the Animal Liberation Front that engages in direct action through the releasing of animals from laboratories, for example, is translating an abolitionist or liberationist commitment to removing animals from situations of human ownership (and thus use) (Best and Nocella 2004). Someone who believes in legal and political rights for animals may seek to work for the Great Ape Project, or support the lobbying efforts of Humane Society International Canada, illustrating a fairly clear and direct link between their analyses and their preferred forms of praxis.

The connections between people's ideas and their political labor are not always tidy or absolute, however. As noted, in practice, theories and ideal types often become less defined and more fluid. Some advocates may also organize their political labor around a specific issue or groups of animals, draw from more than one of the conceptual approaches listed, or eschew theory altogether. Vegans, those who promote cruelty-free living and who do not consume any products derived from animals, may be abolitionists, but there are also vegans who have different political beliefs about animals. Abolitionists may have companion animals in their homes and support rescue organizations and shelters. A worker in a humane society may believe in animal liberation, but opt to pursue welfarist political work because of her geographic context and/or life stage (Gaarder 2011b). It is also not uncommon for an organization to work toward and appreciate small steps or improvements (e.g., a ban on gestation crates), even if their end goals are more ambitious and transformative (such as an end to the consumption of animals for food). All people have inconsistencies and contradictions in their political beliefs, including those working for animals. Moreover, the real world of interspecies political work is complex, limited, contingent, and affected by political economic factors and varying degrees of opportunity (Dave 2014). People make choices for various reasons, and, simultaneously, are constrained in their abilities to make those choices.

At the same time, political consciousness and ambitions can have unexpected or accidental origins. Two Canadians, Steve Jenkins and Derek Walter, thought they were obtaining a "micro" pig to keep in their home, but the new addition turned out to be full-sized. Dubbed Esther "the Wonder Pig," she fundamentally changed Derek and Steve's understanding of pigs, and after researching the issues further, they became vegans and social-media-savvy animal advocates. The two men virtually opened their home to the world so that others could learn more about Esther's personality and intelligence, with the hopes of building greater compassion toward pigs and other animals. Their advocacy work has continued to expand and, in the fall of 2014, thanks to crowd-funding, the Happily Ever Esther Farm Sanctuary was created. Undoubtedly, most people are not going to accidentally become founders of animal sanctuaries, but this story is a reminder that some political ideas and projects are not preplanned, but rather develop due to transformative learning experiences and inspiring interactions with others, including animals.

Still other kinds of work for animals' well-being may not be organized around or stem from a rights orientation. Many people work for wild animals through conservation initiatives, inspired by species-specific or broader ecological concerns. Consequently, work may be done to preserve and protect animals, locally or across borders. In some cases, individual animals' lives may be forfeited in pursuit of a broader commitment to natural resource management or ecosystem preservation. In these cases, divisions can emerge between those interested in protections for animals and those prioritizing the broader environment. Some animal advocates actively organize direct actions against wildlife culls or hunts sanctioned by natural resource bureaus. These sorts of cleavages may revolve around "wild" versus "domesticated" animals, with environment-minded advocates prioritizing the former. Yet this dichotomy may be problematized for a number of reasons, including individual people's own politics or because of the liminality of individual animals and whole species who do not fit tidily into either category (e.g., foxes or mink who are kept in cages on fur farms). Regardless of its intellectual underpinnings, every type of advocacy requires labor.

The Work of Animal Advocacy

Advocacy is an imperfect term, but I have chosen it to try and capture and illuminate a broad range of political work revolving around the active support, promotion, protection, defense, and betterment of self and/or others. In terms of strategies, advocacy may include but is not limited to proactive or responsive efforts, the pursuit of individual or larger levels of protection, educational campaigns and programs, direct lobbying of policy makers, and different forms of coalition-building, organizing, and mobilization. Given the breadth of animal issues, the emphases, vehicles, and routes vary greatly, but all require work and workers. Although this chapter focuses particularly on organizational bases, there are also individuals who have opted to use their skills and/or careers for animal advocacy who should also be acknowledged. They may be self-employed or make space within an established employment sphere for animal-centric tasks, cases, or projects. This includes educators, lawyers, artists, writers, photographers, filmmakers, painters, among others. There are also many people who do unpaid advocacy work in addition to their paid career.

People working for/with animals may do so through networks or groups with specific and/or broader goals (Markovits and Crosby 2014). Although both women and men engage in this kind of political work, animal-centric advocacy of all kinds is numerically dominated by women (Herzog 2007). There are also often gendered divisions in terms of tasks, titles, and labor (Gaarder 2011b). Animal advocates may engage in the work once in a while (such as canvassing or helping to organize an annual fundraiser) or continuously. Examples include a group of women and men who form a rescue focused on fostering and finding homes for dogs; a collective of students promoting veganism and animal liberation; a community-based coalition seeking to close an entertainment facility holding animals. This kind of work is unpaid, and, in fact, the people involved often spend their own earnings on materials or actions. Those involved are driven by their politics and often their passion. The shape, structure, and approaches of these groups vary, and they may have the benefit of localized democratic decision making or develop different informal leadership structures. Autonomous organizations are not dependent on pleasing external over-seers or those who provide funding. They are shaped by and reliant on the degree of commitment and the availability of those involved, and therefore emerge, endure, or dwindle accordingly. Their reach and sphere of influence is affected by their focus, as well as by where they are located.

Some local groups may form or stem from established organizational bases. For example, an international network of Roots and Shoots youth organizations is coordinated through the Jane Goodall Institute, thus groups can receive resources, share information, and fly a recognizable flag, while determining their own local projects. The Jane Goodall Institute itself is an example of the formal, structured organizations where much animal advocacy work is done. Many nongovernmental organizations have been created to do explicitly political work. Others may engage in both frontline service delivery for animals and forms of advocacy. The same or different people can be tasked with these many responsibilities. For example, it is not uncommon for a humane society to offer rescue and adoption services, but also to engage in community-based educational initiatives and perhaps to pursue some lobbying for legislative change, as well. This is particularly true of regional, national, or international societies. Farm Sanctuary in the United States is another example of an organization that does multifaceted work, as it offers refuge for animals, while workers therein also pursue a range of political and educational strategies that target the causes of suffering. Certain veterinarians or veterinary organizations such as Community Veterinary Outreach in Ontario, Canada, provide care exclusively or partially for the animals of poor people within their own communities, and may also engage in broader research or advocacy work related to poverty, homelessness, and social services.

An organization such as the Gorilla Foundation is responsible for daily care and work with Koko and Ndume, gorillas who live onsite in California, but education and other kinds of conservation work are pursued, as well. Koko is a gorilla who communicates through sign language and, as a result, she can reveal her own personality, desires, feelings, and thoughts in a way that humans can easily understand. As a result, Koko has become a kind of ambassador for her species. She participates in broader educational and conservation efforts not only by offering people a glimpse into gorillas' minds and thus fostering understanding and empathy, but also by sharing her own thoughts about how people ought to treat gorillas (Patterson and Linden 1981). Koko has not only been mobilized as a symbol, but also framed (accurately) as an active, multidimensional social agent by Francine Patterson, director of the Gorilla Foundation, and by the other people who work in the organization. At the same time, Koko has been given many outlets through which to express herself as an individual and as an advocate, including with art/painting and through appearances in videos and documentaries. In some, Koko speaks directly to the camera and shares her thoughts about people's behavior toward gorillas. Koko the living being is not an attraction for visitors, however; only special guests or members of the media are invited to the Gorilla Foundation itself for specific purposes.

Many multipurpose or hybrid organizations have different branches and the same or different workers or teams may be responsible for education, research and policy, government relations, and so forth. Those responsible for frontline animal service, such as investigating animal cruelty, can still see their daily work as a kind of individualized advocacy. Depending on the organization, they may also be directly involved in certain types of decidedly political work. These organizations are generally reliant or primarily dependent on donations from people or private organizations. They may or may not be officially incorporated as charities; some jurisdictions like Canada prohibit the issuing of tax receipts for donations if more than 10 percent of an organization's work is considered "political," for example.

Animal advocacy organizations of all kinds have things in common, particularly when it comes to crucial operational work. Clerical work is essential to daily functioning. Other kinds of integral work include office management and/or operations, and research. It is not uncommon for an advocacy organization to forego renting an office to save money, and, instead, workers complete their tasks remotely/at home/on the road. Financial record-keeping, database management, and other foundational administrative tasks are still essential regardless of the physical makeup of an organization. Given that most of the organizations in question rely on donations, fundraising and "donor relations" often require a lot of work and attention. Because online outreach, education, and fundraising has become so central to political work, organizations may have in-house social media staff, web specialists, graphic designers, and/or information technology staff, or they may rely on outside or contract support for these dimensions. Some have lawyers and/or legal researchers and policy analysts, campaign coordinators and organizers, and issue-focused specialists.

The quality of these jobs *as jobs* varies within and across organizations. The essential clerical positions are feminized, as they involve a large number of women, and exemplify the "pink collar" work associated with women's secretarial, administrative, and office management work more broadly. These positions are on the lowest side of the pay scale, but workers may feel a greater sense of satisfaction being employed in animal advocacy organizations, in comparison to certain or even many types of businesses. This can be true across positions. While many of the tasks remain the same, the goals and outcomes can be seen as more worthwhile and laudable, thus workers may feel more proud of what they do and the organization to which they contribute. As a result, the factors that shape perceptions of job quality are likely broader than the material conditions of work alone. Some workers find no problem with dirty work, for example, and/or actively participate in their own unpaid overtime and other forms of sacrifice, out of a commitment to their work and the animals they help (see, e.g., Bunderson and Thompson 2009).

At the same time, those who run or lead organizations that promote social justice projects have been critiqued for expecting a disproportionate level of commitment and martyrdom from their own employees, and exploiting their emotional motivation and passion for the issues (see, e.g., Franzway 2000). A tension can emerge between those who feel that animal advocacy is a movement and a cause that deserves substantial personal sacrifice and dedication, and those seeking (or needing) greater work-life balance or remuneration. Of course, even workers who feel strongly about the goals of their work can become burned-out or frustrated over time if their daily work lives are difficult and their pay checks small. Such feelings can lead to high turnover or to decreased morale which, in turn, affects the effectiveness of the organization. In transnational contexts, where international organizations may exploit local peoples, the dynamics of pay and other kinds of tensions become even more complex (Soldikoff 2009). There is a large body of literature, particularly in anthropology and geography, on the politics of international conservation initiatives and how local peoples and their livelihoods are involved, displaced, discounted, or positively affected.

Given the monitoring of organizations receiving donations by third-party assessors (e.g., organizations that "rank" charities) and the increased self-reporting of financial information (like what percentage of funding is allocated for administrative and staffing costs), this tension is not surprising and likely to increase. Some members of the donating public, including major donors, want their money to "go to the animals," and not to the people responsible for doing the work essential to helping the animals and functioning as an organization. This is but one of a number of challenges that emerge when relying on donors for funding. Erratic and unreliable funding that can diminish in recessions or times of higher unemployment and underemployment is also an issue. The politics of donor influence is particularly challenging when wealthy individuals or corporations seek to determine what is done, said, and not done. Organizations that rely heavily on volunteers or other unwaged workers (such as unpaid interns) are often lauded, and while some volunteers remain dedicated for long periods of time, high turnover is common. They may lose interest or become limited in their availability by a need to secure paid work. The volatility of volunteer reliance means more time and labor are needed for their recruiting and training, and can mean longer-serving workers are also required to allocate more energy to supervising newer and less experienced people, which takes time away from their other work.

Some, but very few animal advocacy workers in nongovernmental organizations in countries like Canada are members of unions who collectively bargain their working conditions. Municipally based or run humane societies are more often unionized. Labor relations in unionized NGOs vary. Greenpeace Canada was criticized by worker advocates in 2003 for illegally locking out 13 of its canvassers, for example (Shantz 2012). Unionization rates are very low across small, nonprofit organizations in many countries, thus animal organizations are not unusual in this regard (Baines 2010). In these organizations, employment relations and the conditions of work are handled primarily by directors and also are governed by the organization's board if one exists. Accordingly, workplace rights and benefits such as paid sick days and/or vacation, health care packages or insurance, and pensions may or may not be afforded to workers. If nonprofits are run by boards, their composition can also change over time, leaving permanent staff dealing with well-intentioned but often inexperienced supervisors and employers.

Many animal advocates actually spend very little, if any, time with animals. The reality of much animal advocacy work is that it involves dealing with people: members, donors, employees, coworkers, volunteers, employers, farmers, social media followers, anonymous online trolls, policy makers, lawyers, corporate leaders and spokespeople, journalists, bureaucrats, political staff, politicians, and so forth. Accordingly, interpersonal abilities, emotional labor, emotion work, leadership skills, and other kinds of "soft" skills are essential. Animal advocates engage in banal and monotonous daily tasks, as well as creative generation, negotiations, coalition building, micro- and macro-strategizing, and tactical decision making. Overall, advocacy work is multifaceted and challenging in many ways, including emotionally.

Emotions at Work

In addition to the conditions of work being a possible source of stress, the challenges of advocating for animals and knowing intricate details about how animals are treated can make the emotional work of animal advocacy particularly significant and difficult. Animal advocates are often fuelled by data and "reason-based" motivations along with empathy and compassion (Glasser 2011; Herzog 1993; Taylor 2004). Advocacy work is pursued because people care; they are motivated by a desire to make change and to improve lives. Some positions involve direct interactions and care work for individual animals, including those who have experienced extreme suffering, but emotional dynamics also affect those working "for" animals exclusively in a political sense. Compassion fatigue is a concept used to recognize that people who work with those experiencing pain, neglect, abuse, and/or trauma may themselves experience deep psychological and emotional effects, or what researchers also call secondary traumatic stress disorder (Figley 2002; Jones 2007). Sometimes dubbed the cost of caring, compassion fatigue is prevalent among people who do nursing work, especially in palliative and pediatric care, and the social services more broadly, particularly when confronting the abuse of women and children.

Those who work with/for animals can also suffer from compassion fatigue and/or secondary traumatic stress (Bradshaw, Borchers, and Muller-Paisner 2012; DeMello 2010; Figley and Roop 2006; Rogelberg et al. 2007; Sanders 1999; Taylor 2010). It manifests through burnout, anger, sadness, despair, depression, and/or addictions. The depth and breadth of the harm done to animals in the legal and illegal animal trade, industrialized agricultural contexts, and testing and research facilities can be particularly staggering, as is the scale of the suffering. Enlisting and expanding on Bonnie Smith's (1998) writing, Carol J. Adams (2012) emphasizes the significance of traumatic knowledge in animal advocacy and the continuous reencountering of traumatic experiences. Becoming aware of what happens, how often, and in how many places, is deeply affecting. So, too, is seeing the embeddedness and normalization of social contradictions and violence, people's complicity with or proactive promotion of suffering, and the enormity of the challenge (Fitzgerald and Taylor 2014). Animal politics mean recognizing the unevenness of the power differentials that see small-budgeted nongovernmental organizations and citizens pitted against national and transnational corporate interests with extreme financial resources, as well as social, cultural, political, and legal capital. Pointing to the web of powerful interests normalizing and profiting from animal exploitation, Barbara Noske (1989, 1997) employed the term "animal-industrial complex," and Richard Twine (2012, 23) defines it as a "partly opaque and multiple set of networks and relationships between the corporate sector, governments, and public and private science."

Arnold Arluke and Clinton R. Sanders' (1996) idea of a caring-killing paradox can also intersect with compassion fatigue, particularly in workplaces like shelters that euthanize animals when workers who care, must kill, because of policy, space, and/or budgetary constraints. Similarly, Lori Gruen (2013) posits the concept of "empathetic overload" to recognize the emotional challenges for those who care about animals. These processes have gendered implications. Notably, it is women who do the bulk of animal advocacy work, especially as volunteers (Davis and Lee 2013; Gaarder 2011a, 2011b; Herzog 2007; Kimmerer 2012; Markovits and Queen 2009; Markovits and Crosby 2014; Nuemann 2010; Peek, Bell, and Dunham 1996), thus they are disproportionately affected. Moreover, women are socialized to be particularly caring, empathetic, and compassionate, and this further exacerbates the effects on their emotions. Both individual well-being and advocates' abilities to effectively do work are affected, as people can become bitter, cynical, hostile, and/or withdrawn.

There is widespread awareness of the emotional costs of engaging in both direct care work and concerted political advocacy among the animal protection (broadly conceived) work force and movements. Individuals share strategies and seek to create supportive affective connections, and a number of formal programs have been created with or without associated fees. Such initiatives are important and can help lessen the emotional pain of being aware of and seeing the depth and breadth of the harms inflicted against animals. However, as long as these forms of violence are inflicted, both the animals themselves and the people who empathize will be harmed (in different ways, of course). Some advocates also become frustrated with each other, and the heterogeneity and divisions among animal advocates both reflect and create cleavages. Shelter volunteers may look critically at paid shelter staff required to perform particular tasks, such as euthanizing animals, for example (DeMello 2010). People working to combat factory farming can become angry at those who work daily for cats and dogs or certain wild animals, yet eat other animals and even hold meat-cooking barbecues as fundraisers for their shelters or organizations.

Nevertheless, people who work with and for animals also do emotion work and develop coping strategies for dealing with the intellectual and emotional challenges of their labor. Multiple strategies are used by the same and different people, including the directing of blame (to owners, employers, paid workers, and so on), emotional distancing, moral certainty, compartmentalization, an emphasis on doing the best among a series of poor options, and/or a focus on root causes or accomplishments and victories of all sizes (Arluke 1994; Frommer and Arluke 2010; Taylor 2004). Interactions with animals are also widely identified as a source of inspiration and sustenance for advocates, even those that are emotionally painful like witnessing death. This is particularly true when workers have gone to great lengths to ensure the animal feels joy and has a "good life" while alive, and especially if the animal had experienced pain and suffering for much of her or his life (Hua and Ahuja 2013). These dynamics are powerfully evident in spaces of refuge like sanctuaries (Baur 2008; Laks 2014; Marohn 2012; Westoll 2011). At the same time, workplaces that combine frontline animal care with advocacy also bring distinct emotional challenges and still require people to undertake extensive emotion work and management. In rehabilitation and health care facilities for wild animals, such as the Toronto Wildlife Centre and the Salthaven Wildlife Rehabilitation Centre where people seek to heal injured or diseased individuals, the animals may be cute, sweet, and/or endearing, therefore affectionate feelings easily develop. But because the focus is on returning the patients to their natural habitats, emotional management and distancing is essential. Emotion work is also crucial for handling the fact that many animals cannot be healed or saved.

In her study of a Guatemalan rehabilitation centre for animals who had been taken from the wild to be bought, sold, and held as pets, Rosemary-Claire Collard (2014, 160) identifies an especially difficult process of instilling fear of humans into the animals, one driven by the hope of being able to release and rewild them. She writes, "I refer to these practices as misanthropic because they are designed to instill in the animals fear and distrust, even hatred, of humans. Such techniques include spraying animals if they exhibit 'unnatural' behaviors (e.g., approaching the floor too often, for monkeys; or coming too close to humans), not speaking in front of parrots (if any parrots learn to speak, even *hola*, they are not released because their learned speech might interfere with wild parrot calls), being stern to and distant from animals, and avoiding touch." While the intentions and rationale for such actions are clear, for workers motivated by a love for animals, withholding affection and fostering fear and hate is nevertheless emotionally complex and trying.

Workers at sanctuaries—places where animals who usually have been saved from situations of abuse, neglect, or slaughter—engage in a range of processes to decommodify, reanimalize, and/or individualize those in their care, by providing names, uncovering and respecting their personalities and desires, and sometimes reframing the animals as representatives of and advocates for their kind. For example, Brenda Bronfman, founder of the Wishing Well Sanctuary in Ontario, Canada, begins public tours by explaining that the animals who live on site are a mere drop in the ocean when it comes to the animals used and killed in industrialized agriculture. The animals are thus understood to be ambassadors and essential to a process of building understanding and transformation in people. Farm work, dirty work, and care work are performed daily to ensure the animals at the sanctuary are healthy and happy, but a core purpose of the sanctuary is humane education and the igniting of social change. Sanctuaries are differently able to engage in onsite humane education because of their sizes, finances, and the animals' needs, but most now use social media to allow a broader audience to learn about the individuals in their care. Working visits that combine a few hours of unpaid labor with the opportunity to interact with the animals are another strategy used by organizations such as Cedar Row Farm Sanctuary.

In the powerful documentary *The Ghosts in Our Machine* (2013), Jo-Anne McArthur's work photographing the often hidden animals (those in industrialized agriculture, slaughterhouses, fur farms, and so on) is highlighted (see also McArthur 2014). She speaks frankly about feeling like "a war photographer" covering a largely invisible war against animals and of having posttraumatic stress disorder as a result of what she has seen. She also reveals that the hardest part of her work is leaving the animals behind because she is not there to liberate them, she is there to photograph them in order to capture and expose the situations into which the animals have been placed. This provides an example of an animal advocate who has eschewed political work that might be more personally rewarding, and instead opted for a route that is more personally damaging because she feels it will be more effective and influential for affecting change. Those interested in working for animals must negotiate not only the realities of the work force, but also the politics of possibility that infuse different vehicles and routes.

Certain types of work for animals require people to move beyond emotional turmoil to physical danger. Rangers in central Africa tasked with patrolling the national parks housing the world's last remaining mountain gorillas, for example, are doing so at great personal risk. "Wildlife rangers endure similar ordeals to soldiers in combat. They routinely face death, injury, or torture from poachers, and the wild animals they protect can kill them too. In the DRC [Democratic Republic of Congo], which has been riven by almost two decades of civil war and political instability, about 150 rangers have been killed in Virunga [National Park] alone since 2004" (Neme 2014, n.p.). Sean Willmore, president of the International Ranger Federation, says two rangers are reported killed somewhere in the world every week but that the number could be twice as high (Ibid.). It is not only poachers but also rebels and militia that endanger conservation workers. Nevertheless, local people still seek out work with conservation organizations because of their love for the animals, and/or because such positions are seen as "good jobs" among the few options available. "Virunga national park, a state institution, is widely seen as an exemplary public employer. Rangers are paid twice what the government allocates them, thanks to donor support, and living conditions are good. Rangers' working hours are long—they often work 24-hour shifts—but they say wages are generally paid on time, while holidays and days off are well regulated, and the park now provides for maternity leave in the rangers' contracts" (Hatcher 2015, n.p.). Indeed, in 2015, women were hired as rangers for the first time.

Some wealthier animal workers from the global north are choosing to pursue work in higher-risk regions to help endangered species. Conservationists, scientists, and veterinarians are going to countries like Rwanda, Uganda, and the Democratic Republic of Congo through organizations like the Canadian Ape Alliance, Gorilla Doctors, the Dian Fossey Gorilla Fund, and Docs 4 Great Apes, among others, to engage in political economic development with local communities, and/or to deliver primary health care to the people and even to gorillas themselves. The organizations share a commitment to the apes, but have different focuses, approaches, and emphases shaped by the agency, skills, and politics of those involved. Most work alongside local veterinarians and advocates, and have proactively chosen to offer their labor and skills, sometimes without charge. They reveal some workers' desire to expand daily care work to a broader level and reach, and to engage in a decidedly political form of work simultaneously. Organizations like Veterinarians without Borders and medical branches of transnational nongovernmental organizations like the World Animal Protection (formerly the World Society for the Prevention of Cruelty to Animals) do similar work in poor communities, after natural disasters, or during wars around the world. The commitment and courage of those who enter crisis zones and dire, dangerous situations precisely to serve and save other beings within and across species is noteworthy.

Working for Working Animals

Depending on how animals' work is defined, as noted in chapter 2, one could argue that a number of organizations are advocating for animals' right to engage in subsistence and care work free from human interference and violence. These organizations do not generally frame their work or animals in these precise terms, however. Yet there are some advocacy organizations that explicit say they work or have worked for working animals.

As outlined, many animal advocacy organizations were founded precisely because of cruelty toward working horses. The Royal Society for the Prevention of Cruelty to Animals (n.d., n.p.) highlights campaigns for pets, farm animals, wildlife, and animals used in research as their contemporary focuses, and explains both its history and evolution in this way: "When we were founded, our focus was working animals, such as 'pit ponies,' who were worked down the coal mines. But we've changed with the times. During the First and Second World Wars we worked to help the millions of animals enlisted to serve alongside British, Commonwealth and Allied forces. And, our work with pets that we're best known for today, only developed with the trend to keep them." The visible, manual work of pulling is widely considered animals' "work" and equid-power has largely been abandoned in the global north except in specific ethnocultural communities (such as regions with large Amish or Mennonite populations), in spaces of tourism/entertainment, and on certain farms wishing to retain or revive this historical practice. There are campaigns and organizations working to end the current use of carriage horses in places like New York City and Sacramento, for example. There are a small number of organizations that claim to advocate for people who employ service dogs and the dogs themselves, but their work appears to focus on the former. Interestingly, Vegans International Voice for Animals has a campaign called White Lies, which focuses on cows whose milk is used for human consumption. For Mother's Day in 2014, the organization launched a campaign framing cows as "Britain's hardest working mums" because they are continuously impregnated and required to produce milk which is taken for people's use (Viva! Health 2014, n.p.). World Horse Welfare, also based in Britain, works locally and internationally to improve horses' lives both during and after their formal working lives through research, political campaigning, and frontline programs.

A number of organizations focus on animals' lives after their careers have ended. This emphasis has long been part of animal advocacy work and women's organizations set up the first retirement homes for draft horses (Beers 2006). In some cases today, organizations work to prepare animals for "second careers." This is most apparent with former race horses who are socialized to make them more suitable for adoption by leisure riders or amateurs who wish to do show jumping, dressage, and so forth. In other cases, the animals are taken into sanctuary for retirement (such as at the Performing Animal Welfare Society in California), or homes are found with people who will provide care until the animal's death.

Much of the advocacy work concentrating on "working animals" focuses on the global south where millions of horses, donkey, mules, camels, elephants, and other animals continue to do pulling and trekking work. There are at least 112 million equids alone working in poor countries (The Brooke 2014, 6). As noted in chapter 2, animals doing pulling work may be in situations of tourism and entertainment or doing labor for individuals and/or families. These animals are often required to do very difficult and tiring physical work, in treacherous conditions and weather, and while wearing harnesses and other equipment that are ill-fitting. They may lack sufficient water, food, and rest (Geiger and Hovorka 2014; Lochi et al. 2012; Swann 2006; Wade 2014). Right Tourism is a British charity that encourages "responsible, informed, guilt-free, and humane tourism" and provides information about elephants, camels, and equine welfare issues in tourist attractions. World Animal Protection (n.d., n.p.) has a broader mandate but also pursues programs focusing specifically on working horses, including in the West Bank, where, along with the Palestine Wildlife Society, medical care and welfare education are provided. There are also campaigns promoting both frontline work and legislative change in Colombia and Thailand, among other countries.

Organizations like The Brooke, Animal Aid Abroad, Animal Care in Egypt, Animal Welfare of Luxor, and the Society for the Protection of Animals Working Abroad (SPANA) focus entirely or primarily on "working" animals. With roots or bases in the global north, these are nongovernmental organizations that provide welfare education, veterinary care and intervention, as well as emergency assistance with feed and water for the people who own or work with the animals, particularly in times of drought. These organizations usually have some paid staff in their home countries and in the field, but also involve volunteers, including veterinarians and veterinary students. Some receive support from high-profile people like celebrities or even royalty. All rely on fundraising. The Brooke's roots trace back to World War I and the plight of horses used in battle, but its efforts have remained squarely focused on working animals. Today the organization engages in both frontline service delivery in countries like Afghanistan, Pakistan, and Nicaragua, and borderless educational and fundraising campaigns. It also pursues data collection and research to bolster its education work and its efforts to secure legislative change. A handful of smaller organizations have grown in countries of the global south like Animal Rahat in India, and there are always individuals who voluntarily work on behalf of local working animals in different ways.

In most of these organizations' efforts, people who use and/or work with the animals are not demonized, but rather the role that animals' labor plays in the lives of poor people is recognized and emphasized. Most of the organizations seem to be heeding critiques about foreigners and patronizing aid programs, thus integrating participatory and dialogic strategies that engage local peoples and validate their contexts, while recognizing gaps in their knowledge of animal welfare. Aklilu Menberu lives and works with donkeys in a city in Ethiopia. In this area it was common to overwork donkeys so their life expectancy was a startlingly low 2 years. Sick or injured donkeys were often released to fend for themselves, no doubt leading to brutal suffering and death. "Before the Brooke, we used to load [the donkeys] 6–7 times a day, 3–5 km for each round which [meant] approximately 18–30km [of] travelling per day, carrying 50–60 kg of stone," Menberu explains. "There was no tradition of feeding or watering them during the day, [and] they were just put out on overgrazed pasture at dusk, where they were exposed to hyena bites ... But these are all history now. We have learned how to treat wounds with water, salt and Vaseline which are available locally. We also use saddles that help prevent wounds and feed the donkeys properly with wheat bran treated with edible oil... [N]ow we know the symptoms before they get sick, we treat them well and give them the rest they need" (The Brooke n.d., n.p.). He and others also learned to cultivate different forage seeds just for donkeys. The Brooke's immediate goal is to ensure that animals are cared for as effectively as possible given the financial constraints of the people, and the larger vision is for policy changes that benefit both people and animals.

In this spirit, in 2013, The Brooke undertook research with women in Pakistan, India, Ethiopia, and Kenya to solicit their views on the role working equids play in their lives. As noted in chapter 1, about two-thirds of the poor people who work with working animals in the global south are women. The research findings illustrate "the extent to which women rely on working equine animals for support in fulfilling their many roles within the household and the wider community. This includes help with domestic drudgery, providing an income for women and their families, and enabling savings by providing transport for goods, water, firewood, animal feed, manure and other produce. Their role also extends to the social sphere of women's lives, as they raise women's status in the community and provide them with opportunities to make their voices heard and to access loan and business opportunities" (The Brooke 2014, 6).

The women's own words about the donkeys are instructive and compelling. "I am lacking words to fully explain how grateful I am and how really my life depends on donkeys," said Faith Wamalwa Kinyua, a 29-year-old woman from Kenya (The Brooke 2014, 33). Lucy Waititu, also from Kenya, explained her connection this way:

> If my baby could speak, she would tell her life as a child of a donkey. The maternity fees I paid while I was pregnant came from income brought by my donkey. When I delivered my daughter, I was able to pay for the Statutory National Health Insurance Fund through money earned by my donkey, which catered for all the delivery fees. My child eats, dresses and lives off income from my donkey... I eat, drink, dress, live off the donkey and more so as a woman and one not employed [in waged work], I work hand in hand with the donkey. Basically the donkey is like me but to plainly put it, the donkey is me. (The Brooke 2014, 40)

Particularly as long as the global distribution of wealth, work, and power remains in its current inequitable structure, there is great need for the well-intentioned but privileged to pay close attention to the webs of human and animal well-being in poor communities.

Animals and Labor Unions

Labor unions are the primary vehicles human workers can use to advocate for themselves and other workers. They are advocacy groups, part of civil society, and not part of government, but are not generally called "nongovernmental organizations." Formal labor unions are a distinct type of advocacy organization with specific legal powers and responsibilities. Workers created unions to gain a collective voice at work, a formal process for defending and advancing their rights, and a means to protect and promote the interests not only of individual workers, but also of working-class people more broadly. Local executives are elected directly by workers; workplace-rooted stewards serve as point people, bridges, and/

or leaders; and various union committees are usually created to unite workers with shared identities or to focus on specific issues. Regional and national leaders are also elected by union members, usually at conventions. Unions as organizations hire some paid staff to provide specific services and work for the members, as well. Both elected representatives and paid staff usually collaborate to provide services to members, including grievance and legal support, educational and training opportunities, and a sense of community. All of the work done by unions is funded by workers directly through their dues, and what work is done is driven by unions' internal democratic processes.

Collective bargaining is a central means through which workers self-advocate and shape the conditions of their own work-lives and workplaces. Workers and union-hired negotiators are tasked with negotiating the best possible contract in the collective bargaining process with representatives of the employer. Although workers almost always have less financial capital than their employers, workers have other kinds of power. These include their labor (and its withdrawal), the potential for generating larger, popular support, and, ideally, the solidarity of other groups. Through their unions, workers also engage in various kinds of organizing and mobilization, lobbying, and campaigns, to self-advocate and/or to propel social change. Unions are workplace-rooted, but are also linked through shared membership in regional, national, and/or international workers' organizations. Workers use their unions to defend gains and promote progressive change, individually, collectively, and socially.

As such, unions are vehicles for advocacy, but also depend on political work. A union provides a structure, but then paid staff and especially members must engage in forms of political action to further advance their interests. Unions are vehicles, and where they go, in what way, and at what speed, depends on the drivers— the members. Unions are heterogeneous, and the specific services, political orientation, campaigns, and other emphases all vary depending on workers themselves. Unions are shaped by the quality, commitment, debates, degree of engagement, attitudes, and interests of those involved at workplace, local, regional, and larger levels. Through political work, lousy jobs that paid poverty wages were transformed into better paying, safer positions. Workers of all kinds—from nurses to retail workers to fire fighters to engineers to professors—have unionized in order to gain the benefits of collective action and advocacy. In Canada, about one in three workers are union members. The level of union membership varies greatly across countries. For example, unionization rates in the United States are much lower at around 11 percent, while a majority of workers in Scandinavian countries are union members.

People who work with animals have unionized for different reasons. In the early 1900s, the Team Drivers International Union (comprised of small t teamsters—those who drove teams of horses) united with the Teamsters National Union of America (primarily cargo handlers) and formed the International Brotherhood of Teamsters, known simply, even today, as the Teamsters. In other words, the Teamsters union's roots are literally with teamsters and a horse's head continues to grace its logo. A number of historians have studied the Teamsters union from different angles, but Clay McShane and Joel A. Tarr (2007) provide a multispecies history, one that illuminates how the human workers and their union understood and approached the horses with whom the men (and it was overwhelmingly if not entirely men) worked. In the early twentieth century, working horses were ubiquitous in cities around the world and horses' labor was essential (see, e.g., Greene 2008). Individual teamsters treated horses differently and there were examples of both cruelty and kindness. McShane and Tarr (2007) argue that teamsters' treatment of horses was influenced by individual workers' agency and beliefs, and particularly by employers' demands. In general, horses were given a day of rest on Sundays, but the human workers were required to do the daily care and stable work on those days. The Teamsters union included suggestions about kinder horse training strategies and options for improved equine welfare in its newsletters. When engaging in political action like strikes, the men demonstrated respect for horses. Feed trucks were allowed through picket lines, and when replacement workers or scabs were used, the strikers chased the men away, while the horses were always returned safely to the barn. Notably, in at least one strike, the Teamsters included a demand for the horses' feed to be increased (it is unclear whether this demand was met) (McShane and Tarr 2007; Teamsters n.d.).

At the same time, as noted, people's visible cruelty toward working horses was one of the key catalysts that inspired much early animal advocacy work. Tensions between modern-day Teamsters who still drive the horse-drawn carriages in Central Park and animal rights advocates continue. While class divisions may not be the central source of disagreement between unions and animal advocates today, many of whom are working or middle class themselves, there continue to be clearly classed—and gendered and racialized—dimensions to animal politics (Einwohner 1999; Gaarder 2011b; Harper 2009). It is disproportionately working-class people who are employed in the industries that involve animals, including those that exploit and kill animals. This fact does not make specific practices more acceptable, but it does mean a classed and intersectional analysis should be brought to bear, and I will further comment on this dynamic in the conclusion.

Instead of formal unions, some people who work with animals have instead opted for voluntary associations, particularly in industries like horse racing and veterinary medicine and in rural spaces (see, e.g., Brooke-Holmes and Calamatta 2014). As noted, some agricultural workers are legally prevented from forming or joining formal unions, as well. Where possible, some farmers and/or farm workers have organized unions, and others have created organizations that may be called a union, but that are different from the formal trade unions discussed. Farmers' groups themselves vary a great deal in terms of their scope and political activities (see, e.g., Araiza 2013; Desmarais 2007; Ganz 2009; Heller 2013; Shaw 2008). Overall, these kinds of worker organizations may provide select services like disability insurance, education, and/or occupation-related advocacy. They are not legally empowered to collectively bargain and enforce the terms and conditions of work, however. They may also lack the financial stability and robustness that stem from regular and universal dues collection.

In assessing the work done with animals, overall, today there are not many unionized workplaces. As noted in chapter 1, with the exception of the higher status "professional" positions, many forms of human work with animals are precarious and not well paid. The reasons for lower rates of unionization relate to some specifics of animal work and some broader trends in labor organizing (or its absence). Many animal workplaces are located in rural communities, often seen as resistant to unions, for example (and some certainly are). People may also be employed in small firms with only a few workers, and unionization is still more prevalent in mid-sized and larger workplaces across sectors. Similarly, many interspecies workplaces are predominantly staffed by women, and particularly in an historical context, a number of unions have had challenges organizing feminized workplaces (Briskin and McDermott 1993; Coulter 2014b; Enstad 1999; Sangster 2010). Some animal workers themselves eschew unionization and its related forms of collective action because of their love for animals and their belief that self-advocacy would "harm" the animals (Miller 2008, 2013b).

Quite a few city-based humane societies in Canada, from Prince Edward Island to northern British Columbia, are unionized, however, with both public and private sector unions but especially the Canadian Union of Public Employees (CUPE). These groups of workers are often part of union locals comprised of city and municipal workers of different kinds. Zoo workers, including those who work directly with animals in positions like keeper, veterinary technician, and animal care attendant are unionized in some places. Certain race tracks and other animal-linked facilities also have groups of workers who may be unionized. For example, some administrative employees of the Ontario Racing Commission, racing judges, stewards, veterinarians, veterinary clerks, and licensing agents are members of the Association of Management, Administrative and Professional Crown Employees of Ontario (AMAPCEO). Workers in natural resource management, certain parks, and animal research facilities on university campuses also have a higher likelihood of union membership due to their location in the public sector which has higher unionization rates more broadly.

Interestingly, in 2014 alone, two different groups of animal workers in Ontario, one in a veterinary clinic and another in a humane society, joined UFCW Canada (the United Food and Commercial Workers union). This is a union that has many members who technically work with animals, but in a very different way. In Canada, the majority of workers who are unionized and who work in slaughterhouses (also called meat packing and "food processing" plants) are members of UFCW Canada. Sites where animals are killed and processed have been among the most heavily unionized of the workplaces that contain animals. Chicago cattle butchers organized a union as far back as 1878 and strikes were widespread (Stull and Broadway 2013). As Amy Fitzgerald (2010) notes, slaughterhouses initially became centralized in major urban centers, and these large, dirty workplaces with miserable conditions were ripe for union organizing. Workers sought to gain a bigger piece of the pie, and union strategies that effectively targeted male-dominated, industrialized workplaces were successful (Brueggemann and Brown 2003):

> During the first two-thirds of the twentieth century[,] labor unions became increasingly powerful in slaughterhouses, even as unions in other industries suffered. Beginning in the 1930s, the United Packinghouse Workers of America (UPWA) and the Amalgamated Meat Cutters (AMC) worked hard to unionize slaughterhouse employees. Reportedly, by the early 1960s these two unions represented more than 95% of the slaughterhouse employees outside of the southern states…As a result, meatpacking became one of the best-paid industrial occupations. (Fitzgerald 2010, 61)

In fact, in 1960, meatpacking wages were 15 percent above the US average in manufacturing (Stull, Broadway, and Griffith 1995).
Yet as a result of corporate restructuring and consolidation, technological changes and the "deskilling" of slaughter work, right-wing legislative changes like "right-to-work" laws (policies designed to weaken unions by making dues payment voluntary while still requiring unions to represent and advocate for all workers), and the movement of US slaughterhouse plants to southern states, unionization rates have decreased. Yet the UFCW in the United States says it still represents 60 percent of workers in the beef slaughterhouses and about 72 percent of pork slaughterhouse employees nationwide (Lyderson 2011). Nevertheless, average wages are low, at just over $12 per hour, according to the US Bureau of Labor Statistics (2014). In plants where chickens are slaughtered, injury and illness rates are higher than manufacturing averages. In plants where larger animals like cows and pigs are killed and processed, the injury rates are double the manufacturing average (Stull and Broadway 2013, 101). As noted in chapter 1, white, US-born male workers have

increasingly left jobs in these facilities, and now racialized workers, women, immigrants, migrant workers, and/or undocumented people heavily populate slaughterhouse workforces.

With the exception of "right-to-work" laws, Canada's history has been quite similar to, as well as affected by, the US context (MacLachlan 2001). As Anne Forrest (1989) explains, for four decades in the mid-twentieth century, collective bargaining for meat-packers established national standards for the work and workers, something that was atypical across Canadian industries and workplaces. Today, not all slaughterhouses are unionized in Canada, but there is noteworthy union density across provinces. It is an uncomfortable irony that so many of those who are unionized and who thus gain a clear vehicle for worker-advocacy are tasked with the work of killing animals. Even more complex is that in the same union there are people responsible for healing and sheltering animals, as well as workers tasked with assembly-line killing of other animals.

As illustrated by the early Teamsters, some unionized workers have used their organizations and collective power to try and improve conditions for the animals with and/or for whom they work. Some contemporary animal worker unions promote the work their members do for animals. For example, CUPE 1600 (n.d. n.p), the union for Toronto Zoo workers, highlights its members roles in preparing nutritional food for animals across North America, and in overseeing captive breeding programs for endangered species. Yet workers across sectors do not often attempt to bargain for better conditions for the animals under their care. People may believe that because the organizations for which they work have mandates of animal protection and advocacy, their union is a specific vehicle intended to address their conditions and lives as workers. Similarly, they may see that the labor performed, the specific tasks required, and the initiatives undertaken are the purview of their employers and/or them as autonomous workers. In their research on zoo workers, Bunderson and Thompson (2009, 43) found that while the workers' "sense of [their work as a] calling may lead to a grudging acceptance of perceived mistreatment by management, it also makes zookeepers less accepting of perceived mistreatment of the animals due to management's action or inaction." In this case, the workers felt that both they and the organizations for which they worked had a moral duty to the animals.

Some unionized workers have sought to directly link people's working conditions to the well-being of the animals under their care. This is a strategy that has also been used by different groups of human-focused care workers like nurses, education workers (e.g., "our working conditions are your learning conditions"), among others, and illustrates what Linda Briskin (2013) calls the "politicization of caring." The argument is that if there are sufficient numbers of workers who feel respected and have manageable workloads, they will be able to provide higher quality care. In some cases, workers may need to engage in more militant forms of collective action to that end. For example, at the Lincoln County Humane Society in Ontario, workers felt compelled to withdraw their labor through a strike as a way of improving their conditions, the lives of the animals, and the management of the shelter itself. During the strike, a sign was made and erected that said "Treat Animals and Workers with Respect."

Overall, the political work done through unions has focused predominantly on improving the jobs themselves and on human workers. Moreover, advocates for animals and advocates for human workers who work with/for animals may be at opposite ends of key political issues. Zoos are a clear illustration of this dynamic. Anti-captivity activists seek to close sites like zoos, while the workers therein and their unions seek to improve people's work-lives in such spaces, thereby potentially creating better experiences for the animals under human care as a result, but not questioning the ethics of captivity. Slaughterhouses are another significant example. The battle over horse-drawn carriages in New York City certainly also illustrates these differences. These flash points expose oppositional positions and contradictions that may be difficult to reconcile. Yet there are different, possible paths forward that interweave human and animal well-being and reconceptualize political work involving animals. I will discuss these dynamics in more detail and present further food for thought—and action—in the conclusion.

The Public Politics of Animals

By exploring the political work done with/for animals, it becomes clear that at the heart of the matter are questions about who is included in political and other communities, in "the social," and indeed in visions of social justice. Political theorists, philosophers, ethicists, and other scholars are tackling the conceptual implications and possibilities of thinking about animals in these ways, and doing so from a range of epistemological and ideological perspectives, particular liberal and Marxian traditions (see, e.g., Benton 1993; Donaldson and Kymlicka 2011; Nussbaum 2006; Singer 2006; Smith 2012). At a practical level, these questions also prompt consideration of the public sector, as both a space of policy and law making, and as an employment sphere. Work is being done—and more could be done—in and through the public sector to advocate for animals.

As noted in the introduction, capitalism is the dominant economic system globally, but not all spaces of work are in the private sector (figures 3.1 and 3.2). Notably, not all work performed within the private sector is for-profit either, and the nongovernmental organizations highlighted earlier in this chapter reflect what is called the third sector, the nonprofit sector, and sometimes the only partially accurate voluntary sector. Also, as noted, although unions are not commonly referred to as "nongovernmental organizations," they are autonomous organizations that are not part of government either.

The public sector is closely linked to government and to the political workers therein—that is, those empowered to make laws,

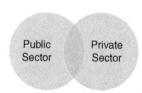

Figure 3.1   Public and Private Sector Intersections

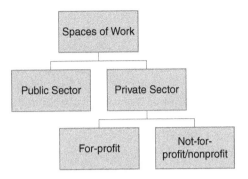

Figure 3.2   Public and Private Sector Structure

regulations, and policies, and allocate and redistribute public financial resources. All governments collect tax revenue and make laws and policies; how much money is collected, on what/who it is spent, and for what reasons vary depending on the particular politics of the governing party (or parties in the case of coalition or minority governments). The public sector can also be understood as "the state," which includes nonpartisan civil servants in governmental ministries, and permanent institutions funded by public revenue and governed by laws and policies, but more autonomously run and managed at the local level (such as the police, schools, etc.). Decisions made in public, state institutions are still funded by public resources, however, and shaped to varying degrees by elected, governmental decisions. Public sectors around the world are funded predominantly through collective, public revenue (taxes), and services are intended to be delivered for the public good. In other words, the goal is not to make money, but rather to serve society. People work in governments to make policies and spending decisions that represent and reflect the priorities of the government (and, in theory, citizens). These policies and spending decisions then affect public institutions—and the work done therein.

There are different kinds of overlap between the public and private sectors, including through the taxes that are collected (or not collected) from the private sector. Nonprofit, nongovernmental organizations also may apply for small pots of public funding distributed in the form of grants for particular projects or dimensions of their work. The political work of governments also affects the private sector through the laws and regulations that are established to curb undesirable acts, encourage positive practices, and establish acceptable standards. Thus the work of policy and law making is interwoven with work done outside of governmental spaces but in the broader public sector. For example, publicly funded inspectors are tasked with investigating workplaces of all kinds to ensure compliance with established laws and regulations. In most countries, public money is already spent on animals through local animal management offices that focus on strays, natural resource or wildlife departments, subsidies to industries like fur farming, seal hunting, or agriculture, among other routes. Some programs that task inmates with working with farm animals or with training service animals receive most or at least some of their funding from public sources.

The public sector intersects directly with advocacy work and animal politics. Advocacy groups target governmental representatives and public policy in their efforts, and seek to have laws introduced and/or changed. For example, the Canadian Federation of Humane Societies, an umbrella group for organizations across Canada, was formed in 1957 specifically to address the welfare of farm animals as there were no regulations at that time. In 1959, Canada's first law, the Humane Slaughter of Food Animals Act, was introduced to establish protocols for transportation and slaughter (but not on-farm treatment).

Different politicians and political parties have responsively or proactively sought to introduce laws and policies, as well as change those that exist, to better protect and serve animals. What work is done is shaped by the political cultures of particular regions, countries, and parties, the power and effectiveness of movements and advocates, the active construction of demand, unexpected events, among other factors, and there are significant differences across countries (Deemer and Lobao 2011; Evans 2010; Lerner, Algers, Gunnarsson, and Nordgren 2013). Current Canadian laws for the transport of animals to slaughter, for example, "allow cattle and sheep to be transported for up to 52 hours continuously with no food, water or rest. Pigs, horses and birds can be transported for up to 36 hours. And there is no requirement for animal transporters to have any training on how to handle animals humanely or to drive safely with them on board. In comparison, in the European Union, most species are not permitted to be transported for longer than 8 hours, unless transporters meet several conditions that preserve animal welfare on longer trips" (Canadian Federation of Humane Societies "Transportation" n.d., n.p.). At present, codes of practice for on-farm treatment are also primarily voluntary, and the existing laws governing transportation and slaughter are seen by many animal advocates as antiquated (Bisgould, King, and Stopford 2001; Francois 2009; World Society for the Protection of Animals 2010).

Citizens and movements may succeed in galvanizing support for a particular animal or issue based on concerted effort and multi-pronged strategies, a local or regional connection or affinity, and/or the securing of support from a key political figure in government. While governments are empowered to lead, many have to be pushed, and even those that are sympathetic or committed to being proactive, both seek and need support from citizens and "stake-holders." Different advocacy groups working on a single issue enlist different strategies shaped by their political and tactical orientation, with varying degrees of success. In recent years, many local, regional, and national governments have updated their laws and introduced new measures to improve animals' lives. For example, a number of jurisdictions have banned the production, sale, and/or import of "foie gras," the extremely enlarged duck or goose liver that is created through force feeding. Some countries in the European Union, Norway, Brazil, Israel, and India have all implemented bans on the sale and/or import of cosmetics tested on animals. France and the province of Québec (often seen by advocates as having especially weak animal protection laws) have both begun developing legislation to enshrine animals as sentient beings, something done in the European Union in 2009 and in New Zealand in 2015. First in Zurich, Switzerland, and then in New York City, a publicly funded attorney has been appointed to serve as an animal advocate. A referendum question sought to extend this practice to cantons across Switzerland but was defeated (Smith 2012).

Political parties have positions on animals that are both predictable and unexpected, and the conventional left-right spectrum is not a clear predictor of animal politics. Green parties are not necessarily the most progressive on animal issues, either. How animal politics intersect with a party's positions on/for people and on the environment has long been inconsistent and uneven. In some countries like Canada, conservative parties are resistant to policies that would curb the actions of their supporters on farms or in corporations, while greater protections for police dogs can fit comfortably with their "law and order" agenda. Certain social democratic parties are keen to use the power of the public sector to regulate and curtail harmful practices, while others, especially those seeking to be seen as noninterventionist, are less proactive.

At the same time, social democratic-rooted ideas of proactively using the public purse to better animals' lives are evident around the world. Scandinavian and Nordic countries, and particularly Sweden, have, to differing degrees, demonstrated a willingness to use the power of government regulation, law-making, and public investment to improve some animals' welfare, and restrict the behavior of animal-using industries, even when the requirements become cost-prohibitive for the businesses and cause bankruptcies (e.g., fox and chinchilla farming) (Levenson 2011). The Swedish Board of Agriculture is empowered to make detailed regulatory changes that affect the treatment of animals across the country. Rules that prohibit the keeping of a solitary horse (since they are herd animals) and the tying of horses in stalls, and that mandate six hours of time with other horses outdoors are in place, for example.

In the city of Athens in Greece, the municipality acts as a sort of guardian for stray dogs, seeking homes for them in some instances, but more often respecting their right to live where they choose. Dogs are sterilized, vaccinated, provided a collar that lists the number of the animal care branch of the municipal government, and then released. The city explains their actions in this way: "Stray animals are an inextricable element of our city and the Municipal Authority is determined to protect them. The City of Athens is one of the few municipalities in the country which deals with the stray animal phenomenon by implementing specific measures which involve substantial financial and human resources" (City of Athens n.d., n.p.) In other words, public money is spent on these programs, and people are employed to do this work. This policy has been in place since 2003 thus endured during times of extreme austerity.

Different municipalities and larger regions around the world have or are beginning publicly funded spay and neuter programs for feral animals, and/or for the animals of low-income people. In 2014, the municipality of New Delhi in India announced a plan to provide some of the city's many stray dogs with veterinary care and assess their potential for becoming service or security dogs.

Leftist governments of different kinds (i.e., those that believe in economic democracy), have had a broad range of positions on animals, but some have introduced interesting measures and used public resources for animal betterment. Cuba is a mixed example when it comes to animals, but particularly notable is that veterinary care is largely without cost, as veterinary medicine is socialized/public, and veterinarians receive their salaries from the state. The socialist government of Bolivia banned the use of animals in circuses. The socialist government of Venezuela has also begun to expand its vision of revolutionary social change to animals and to encourage political work that crosses species boundaries. Named after the dog of anticolonial hero Simon Bolivar, *Misión Nevado* (n.d., n.p.) is a project "rooted in the animal movement and the ecosocialist movement which seeks to integrate the inclusion of animal rights and mother earth's rights into the ethic of new men and women." It is pursued for ethical, health, ecosocialist, and social liberation reasons, and the program's official slogan is that the capacity for love is infinite. With public financial support, community collectives engage in local education campaigns about issues like bull running and cock fighting, as well as deliver services for homeless dogs and cats.

This very brief survey has only captured some of the publicly rooted routes to animal well-being that are being pursued globally today. Yet all offer examples of how those who work in government and in the broader public sector can envision and be engaged in advocacy work with/for animals. The public sector should be understood as not only providing laws and regulations, but also their enforcement. Moreover, public policies both reflect societal priorities, and can establish new, higher standards, particularly on what is condoned and condemned. Thus, the public sector can and should be a space of possibility, through law and program creation, development, and expansion (Silverstein 1996). Programs delivered through the public sphere may be targeted at issues or areas with particular need, or they may be universal, ensuring equitable access regardless of geographic location or factors like income. Public funding for animal-centered or multispecies initiatives does not have the volatility of fundraising reliance, but it is affected by the political orientation and political will of governmental leadership. Notably, all of these dynamics suggest there is great need for an expanded discussion of what it means to recognize that we all live in multispecies societies.

Overall, along with nongovernmental organizations and labor unions, the public sector has an important role to play. Without question, each area outlined in this chapter could be more deeply studied and elucidated, and the specific emphases and contradictions of the political projects and advocacy vehicles identified have been examined in more detail by the scholars referenced, among others. What is particularly noteworthy for the purpose of this book is that people are working in each sphere of political action to reduce or even end suffering and improve the lives of people and animals, but there is still much more to do. It is to these challenges and possibilities, both intellectual and political, to which I now turn.

Anifesto: The Promise of Interspecies Solidarity

A broad cross-section of material and ideas has been introduced in this book to highlight the many intersections of animals and work, encourage greater dialogue, and posit nuanced ways of conceptualizing the social actors, relations, and labor processes involved. I have proposed and elucidated the concept of animal work as both an umbrella term and as a springboard for thinking about and through the diverse, complex particulars of work done with, by, and for animals. Given the breadth and diversity of the data, tidy and totalizing conclusions are impossible. There are a few noteworthy insights that can be induced, however, and these illuminate patterns of commonality, divergences, challenges, and avenues of possibility.

As noted in chapter 2, the data suggest there is no need to entirely reinvent the wheel in order to foster deeper understanding of animals' work. By using a theoretical tool kit that draws especially from gendered labor process approaches and feminist political economy, and that enlists newer ideas like the continuum of enjoyment and suffering, we can uncover and thoughtfully understand the range of work done by animals. At the same time, the evidence makes clear that we also need to move beyond existing theoretical frameworks, by developing animal-centric ideas of body work, for example. In other words, in this volume, I have identified, assembled, and adjusted valuable, relevant existing terms and frameworks, proposed an expanded conceptual vocabulary and vision, and suggested areas where more intellectual work needs to be done.

This book is propelled by concerns that are both intellectual and political, and how we understand can and ought to be connected to how we act. The local-global evidence about animals' lives contains inspiring and novel examples of how people work for animals in interpersonal and political ways, of how people and animals work together with respect and kindness, of animals' diverse and even surprising contributions, and of laudable, moving expressions of compassion within and across species. Yet today's animal work world is marred by more suffering than enjoyment, for both people and animals, particularly in industrialized, for-profit systems. The extremes of the good and the abhorrent, and the spaces in between, are telling. They reveal that people are capable of empathy, of acting in ethical ways, and of challenging both the causes and symptoms of suffering. But they also reveal deep contradictions in how individuals and whole societies view, treat, and position animals, particularly when people are seeking profit or needing to make money, or even just trying to get by with the very basics. It is in spaces of work where the most incomprehensible and destructive things are done to animals—involuntary or deliberately—and where animals are most brutally used then discarded.

Therefore, this book would not be complete without more serious discussion of how to alleviate and/or eliminate suffering and forge paths to better worlds of work. Simply gaining a deeper understanding on its own is insufficient at this historical juncture and would abdicate my responsibility as a labor studies scholar, particularly one working for the public good at a publicly funded university. Members of multiple species who live, work, enjoy, suffer, and die are part of the public I serve. In many ways, connectivity and difference, as well as inclusion and expansion, are central to understanding the present and possible future of animal work.

Connections and Differences

There are some clear connections between types of work performed by people and by animals. Core labor process concepts and theories—anthropocentric approaches intended to be helpful for understanding people's work—are relevant to the study of animals' work because they allow us to effectively capture key dimensions of what is going on. Human and animal workers in a number of sectors engage in comparable labor processes, as illustrated in chapters 1 and 2. These commonalities include what work is done, as well as positive and negative experiences of daily labor and workplace relations. Notably, there are many processes of "shared suffering," as Jocelyne Porcher (2011) puts it. This shared ground is significant, and many animal ecofeminists, neo-Marxists, and other critical animal studies scholars also stress linked forms of alienation and exploitation across species lines (Adams 2010; Adams and Gruen 2014a; Gaard 1993; Kemmerer 2011; Kim 2015; Nibert 2013, 2014; Taylor and Twine 2014). Put concisely, these writers emphasize that similar organizations, ideas, and processes harm and oppress both animals and specific groups of people, particularly women, migrant and racialized workers, and/or working-class people, overall.

At the same time, people and animals' work is not identical, nor are their positions in workplaces. People and animals are differently positioned in relationships and systems of power. Building on Mary Louise Pratt's analyses of colonialism, Rosemary-Claire Collard (2013, 62) argues that multispecies spaces are also "saturated with deeply asymmetrical relations of power, structured by histories, knowledge systems, and political economies that position animals as subordinate to humans…they are mutually constitutive and radically asymmetrical." Such an approach recognizes that both people and animals shape their shared workplace relationships and experiences, but that they do not do so from equal positions. Usually people have more power and their decisions can greatly affect animals' experiences and lives. But animals are not powerless or voiceless, and they shape elements of daily practice in all sorts of ways. Moreover, workplace relations are not always underscored by a hierarchy of human over animal. In some contexts, the dynamics are problematized by animals' socially ascribed status and value, and they are not necessarily viewed as inferior to people. Horses priced at hundreds of thousands or millions of dollars are viewed differently from working-class grooms being paid poverty wages, for example, and the latter group is often seen as much more disposable by owners and/or employers (although the labor of both groups is intensely used). Horses themselves are also ranked differently and hierarchically based on ascribed economic and/or social value (Coulter 2014a). Yet, notably, in most cases, regardless of where horses are positioned in these socially constructed conceptual hierarchies, their lives are still organized primarily to suit human economic purposes.

Much recent animal studies and human-animal relations research has emphasized partnerships, connections, and the dissolution of boundaries between people and animals (albeit often in spaces of leisure, at least for the people). While these dimensions are important and relevant, there are still material and cultural structures shaping the terrain upon which individuals of all species act. Scholars ought not to gloss over or deny various kinds of hierarchies in a fervent pursuit of commonalities and connectivity. Partnerships may be present, but they do not operate in a vacuum. On the other hand, critical analysts should not negate or ignore collaboration, linkages, and dynamics of respect by overemphasizing systemic analysis and imposing a pre-prescribed assumption of human domination. Even if choosing to highlight and prioritize structural and social inequities between species—an understandable emphasis given the state of so many animals' lives—monolithic statements or inaccuracies are not helpful and can lead to being discredited or dismissed. Highlighting and analyzing the truths about people and animals is a delicate but important balancing act, best achieved through a commitment to evidence, reflexivity, and contextualization.

Moreover, rather than generalizing, the intellectually prudent recognize differences and cleavages among people or even groups of people based on gender, race, ethnicity, nation, sexuality, ability, age, and so forth, as well due to people's own subjectivities, politics, and forms of agency. Animals themselves also have things in common and differences. People and animals have things in common, as well as differences. The context and specifics matter. We can identify patterns, but this does not mean we can ignore, deny, or erase counterexample and exceptions.

This challenge is potently illustrated when thinking about animals' work and choices. In most cases, animals are born into or placed into situations of work and not given a choice about their location. They may be able to influence aspects of their daily life and work as examined in chapter 2, but they rarely have the option of opting out entirely. Fittingly, the memorial to animals in war in London, England, is emblazed with the words "they had no choice." This truth has broader applicability. In fact, the connections between work and choice are complex even when thinking about human workers. Some neo-Marxists, for example, continue to use the concept of wage slavery to critique the notion that working-class people choose their jobs and/or are truly free. Certain people are still literally enslaved and taken and traded as property; the International Labor Organization (2012) estimates that about 20 million people are in situations of forced labor, about 55 percent of whom are women and girls. At the same time, labor researchers are further problematizing the dichotomy of free and unfree labor. For example, Siobhán McGrath and Kendra Strauss (2015) argue for understanding both the degrees and forms of unfreedom in labor relations. Since virtually no societies provide guaranteed basic incomes to all citizens, most people have to work for wages for many decades in order to survive; this is not a choice. Even in subsistence societies, people needed to do work in order to survive and sustain themselves and others. Everywhere on earth, people are differently able to make choices about what they do and where they work, but the need for work, and especially income-generating labor, is structured into political economic systems the world over.

However, notably, while human and animal workers are connected, are constrained in their abilities to make choices about/at work, and are harmed by similar practices, their situations are not identical. The reality for animals is normally much more limited and limiting; people still have more choices and control, including over themselves and over animals. Moreover, although often partial and inadequate, more laws, policies, and programs are in place to better protect people at work and after their paid working lives end. In some cases, certain laws exist that govern the conditions of work for animals, such as municipal ordinances for carriage horses that dictate maximum number of work hours per day, a day of rest, prohibitions on working an injured horse, and a maximum age limit. Even in some contexts where animals may not be seen as "working," laws exist, such as the Swedish regulations highlighted in chapter 3 which require that no horse be housed alone or tied in stalls, and that horses have a minimum of six hours of outdoor time with or in view of other horses each day. But, overall, little and/or outdated legal infrastructure governs animals' working lives or what happens to them afterward.

Indeed, as Josephine Donovan (2007, 362) notes, animals are "commodified and quantified in the production process—even more literally so than the proletariat, whose bodies at least are not literally turned into dead consumable objects by the process, though they may be treated as mechanical means." Many kinds of work harm people's bodies and minds, but they do not have parts of their bodies removed as standard practice, and certainly not without anesthesia. Cubicles are cramped, but people can stand up, move around, turn around, and walk away. Even at the most oppressive human workplaces where conditions are undoubtedly awful, such as the manufacturing, assembly, textile, and commodity production facilities found across the global south referred to as "sweatshops," human workers usually go home, have some time of their own, and have at least some potential to unionize and self-advocate. Most significantly, virtually no human workers are put to certain death by their employers and certainly are not born by the billions for a guaranteed, premature ending. Certain occupations and sectors are dangerous, some employers are unscrupulous and uncareful, and soldiers are sent into war zones; there are real risks to human workers, without question. But there are not entire workplaces waiting to receive live humans who will then be killed and processed as standard practice. In response to a public image featuring pigs in a factory farm dreaming of grazing in a pasture outdoors, one woman perfectly captured both the interconnections, and the differences: "I daydream about frolicking outdoors while I'm at my job [too,] but at least my owners don't eat me."

Conceptualizing Animals, Envisioning Change

Thought and action are connected. Social actors, including animals, can be understood in many different ways, simultaneously, even by the same observers and ascribers of names. People within and outside of multispecies workspaces describe and see animals as food, athletes, tools, machines, stupid beasts, worthless subordinates, servants, commodities, sentient commodities, friends, family members, coworkers, partners, employees, as well as alienated workers and/or disposable workers. The conditions of work and larger contextual factors can shape how human workers view the animals with whom they interact, as illustrated throughout chapter 1. So, too, can definitions of work. Jocelyne Porcher has explored how people view the animals with whom they work from a couple of angles (2011, 2012). Her research with farmers reveals that

> farm animals have an important place in work and collaborate with the work. Some farmers think that their animals do effectively work, other farmers think not, reserving real work for equine and bovine draught animals, for example. This perception of animal work also depends on the place of the animal in the production system. For example, a farmer is more inclined to think that a cow works, but a calf does not.
>
> Generally, what farmers say concerning their work relations with animals leads one to assume that they think that their animals work—the words "work" or "job" are frequently employed. However, if we pose the question directly, the answer is frequently "no." It depends on the definition that a given person gives to work. Most frequently a fairly shared notion of work is demonstrated, grounded in constraint, suffering and dependence—a vision anchored in the monotheisms ("we must earn our bread by the sweat of our brow"), more than in a scientific or political definition of work. Thus, if a farmer uses an implicit definition of work based on constraint, and if he [or she] considers that animals' relationship to work are *not* constructed on constraint and suffering, he therefore responds that animals do not work (because they do not suffer). However, if we start again with a different definition of work, based on what the farmer said concerning his animals' relationship with work, the farmer often changes his response. Nevertheless, the place of death in work with farm animals makes animal work more difficult to think about in farming than in other sectors, and with other species of animal. (Porcher 2016, n.p.)

It is true that there are clear differences between requesting or even requiring some kinds of work, and mandating death. This should be kept in mind.

In examining a somewhat unified group of human workers like the early teamsters highlighted by Clay McShane and Joel Tarr (2007), the data also reveal that linked individuals are heterogeneous in how they view animals. The teamsters saw horses as sentient beings, friends, coworkers, living machines, or "fit co-partners in Life's race," intertwined by shared labor and fatigue (McShane and Tarr 2007, 43). Notably, regardless of what people call them, animals *are* friends, family members, allies, supporters, guardians, caregivers, mentors, enemies, survivors, agitators, and countless other identities, including workers.

Therefore, I do not propose replacing the other, multiple identities—and subjectivities—animals possess with the singular category of worker. For one, I have not proposed that recognition of work done by animals automatically translates into an appropriate or accurate application of the term "worker." Yet even in cases where animals may and should be clearly viewed as workers, as is the case for people, this is not all they are. However, there is a need, as Jocelyne Porcher (2014) also argues, to *recognize* animals' work and how much they provide to individuals, communities, and whole societies. I suggest identifying animals' work and their contributions as *another* dimension of their lives as individuals, species, and community members, as a way of thinking more widely and carefully about animals, about people, and about our connections.

What we condemn and condone are affected by what we know, as well as by how we understand it. Seeing animals as working and even as workers may increase their immediate value in particular ways. Becoming "useful" can change how individuals and/or species are seen and treated. Animals viewed as nuisances or dangerous or disposable become ascribed with different meanings and roles (as seen with the rats who detect land mines in parts of Africa, for example). In some instances, this means animals' lives are spared. In some cases, their lives become better; yet in others, their lives worsen. Such a transformation can mean animals become commodities or tools, that they become partners and friends, or some combination. The process of beginning work and/or of being recognized as a source of labor does not guarantee any specific outcome for animals. Being a worker is not necessarily a ticket to a better or worse life for animals: the context matters.

In order to understand as well as recognize animals' work, we need to think about both workplace relationships and larger social contexts. As Donaldson and Kymlicka (2011, 131) suggest, the process of integrating animals into our political communities means rethinking relations and laws, and "thinking about what special abilities animals bring to the mix." Animals bring many invaluable skills, attributes, and possibilities to the present and to possible, alternate shared futures. Thus, here, women's organizing and feminist political economy both offer food for thought. Women, and especially poor women and those from racialized communities, have long fought to have their unpaid caregiving work in homes recognized as work and to have their paid work fairly compensated (see, e.g., Glenn 2010; Little 2007; Naples 1998; Piven and Cloward 1979; Tait 2005). Discursive and cultural obfuscation (or erasure), and economic devaluation have occurred at interpersonal, organizational, and sociopolitical levels. Women's work has been ignored, downplayed, denied, or belittled in interactions with employers and coworkers, by workers' own organizations, and by policy makers. As a result, women have organized and engaged in various forms of collective action in their workplaces, communities, and at a national scale. They have sought to be seen—and seen differently. They have also called for actual, tangible, changes to public policy and law.

Thus, Nancy Fraser (1995, 93) is both reflecting and propelling this dual cultural and material agenda when she argues that "We are currently stuck in the vicious circles of mutually reinforcing cultural and economic subordination...Only by looking to alternative conceptions of redistribution and recognition can we meet the requirements of justice for all." For Fraser, recognition requires a concomitant and interconnected pursuit of redistribution; that is, a different material order without gross economic inequality. Accordingly, she further argues that the concepts of recognition and redistribution can be mobilized not only in the pursuit of affirmation, but for transformation. In other words, these concepts are a vehicle, as well as a goal. They can provide motivation and planks for building a different social vision.

Nancy Fraser was not writing about animals, and animals do not receive monetary pay directly, nor would they be interested in money specifically. But this framework, equally reflected in poor women's organizing and feminist political economy more broadly, nevertheless offers compelling insights that could be enlisted for thinking differently about animals and work. Recognition of animals' work should move beyond noticing and framing their labors, and go even beyond acknowledging the social, economic, cultural, and interpersonal contributions they make. This plays a role, and is sorely needed in many contexts. Yet the conditions of animals'— and many people's—lives require that we confront more than what is going on now. The crucial challenge is to grapple with how we could do better.

Expanding Visions

Work does not automatically mean pain and coercion, nor does it mean pure joy and voluntary involvement. This applies equally to work done with, by, and for animals. Work has a broad constellation of meanings, and how labor is experienced is shaped by the occupations, the work required, the individuals involved, the local and larger context, the socioeconomic system, among other factors. All current societies have determined that at least some paid work is socially essential and (potentially or ideally) even meaningful. Thus, one could argue that if domesticated animals are part of society, they should not be exempt from some expectation of work, from contributing in some way. An argument could be made that requiring work is very different from mandating death, and that the former is acceptable, while the latter is more ethically questionable.

In contrast, some advocates, and particularly abolitionists, would argue that animals are not humans to use, and that animals should be freed from all of their relationships with people. Undoubtedly, there is a compelling ethical core to the argument that no one is a commodity to be owned by anyone else in this world. At the same time, the abolitionist position is both risky and completely safe. It is risky because it is bold, contrary to deeply entrenched sociohistorical ideas, and can be pejoratively called "extreme" or "radical." Yet it is safe because it can free proponents from having to grapple with the complexities of contemporary human-animal relations and the messiness of real politics. Although abolitionists are not completely intellectually uniform and not all would respond in this way, if one's position is that all animals should simply be liberated, then the answer to every question and suggestion for every situation certainly could be the same (and for some it is): total liberation now. We then do not have to recognize or grapple with the fact that millions of people, particularly poor and working-class people, work with animals. Their livelihoods are interwoven with animals, and animal liberation would disproportionately affect these already disadvantaged people in significant ways. A monolithic approach to all human-animal work relationships and work-lives fails to capture this significant fact, as well as what would be lost in this permanent conceptual and literal "othering" of animals, and their expunging from our shared communities. Similarly, the hetereogeneity and the positive dynamics that have been and are present in real spaces of multispecies and interspecies work are also erased. The argument discounts the enjoyment people feel by working with/for animals, that animals have particular skills and contribute in crucial and even irreplaceable ways, and that animals can enjoy work and relationships with their human and/or animal coworkers. Overall, it negates many social and political possibilities and imposes a singular, totalizing agenda onto a more complicated set of relationships, dynamics, and processes.

Accordingly, I will most certainly not defend the status quo, but I also will not propose an end to all human and animal relations. I see ethical, intellectual, and political problems with doing either. There are different ways of thinking through and addressing the many worlds of animal work that take both people's and animals' well-being seriously. I build on existing concepts and projects that hold potential, as well as enlist and propose additional ideas for further intellectual and political work. In lieu of advancing a singular argument or purporting to have the correct, ultimate answer, I offer a collection of interrelated possibilities rooted in the experiences, practices, and politics of the many spaces of animal work, and in the spirit of promise, innovative thinking, and hope.

Toward Interspecies Solidarity

Some animal ethicists and philosophers actively debate different visions for strengthened human-animal relations that recognize animals' capabilities, needs, and desires (see, e.g., Anthony 2009; Cripps 2010; Donaldson and Kymlicka 2011; Fenton 2014; Gruen 2014; Nussbaum 2006; Weisberg n.d.). When animals work for people, their labor is being used by people. Sue Donaldson and Will

Kymlicka (2011) point out that "use" unto itself is not necessarily negative within or across species, however, nor does it need to be. People "use" each other in countless ways and for different reasons, for example, including for support, friendship, companionship, as well as emotional, physical, and financial help. Thus, as part of their broader arguments for new ways of thinking about animals, Donaldson and Kymlicka differentiate between detrimental exploitation, and some acceptable and permissible uses of animals' labor—if embedded within a larger political framework that includes both protective measures and positive entitlements.

Building on these ideas, feminist political economy, and progressive labor strategies, I propose interspecies solidarity as an idea, a goal, a process, an ethical commitment, and a political project. The concept of solidarity is underscored by ideas of empathy. In contrast to sympathy or pity, empathy is about understanding and legitimizing the experiences of others. Solidarity, thus involves support despite differences. Yet it does not mean connectivity or commonality are prohibited. In contrast, as Val Plumwood (2002, 200–2) writes in her call for solidarity with nature, "both continuity with and difference from self can be sources of value and consideration, and both usually play a role." Someone does not need to be "the same" as you in order for you to feel and foster solidarity and for you to stand "with the other in a supportive relationship in the political sense" (Ibid.). When thinking about interspecies relations, undoubtedly this dimension is noteworthy; people do not need to be identical to animals for solidarity to be felt and encouraged. The pursuit of interspecies solidarity involves an expanded sphere of empathy and understanding, but someone could still argue and believe that people are different from animals, simultaneously. Solidarity should be promoted not simply because animals are like us/we are like animals, but because it is the ethical thing to do. Others, whether human or animal, should not have to be like us for us to care about their wellbeing.

Moreover, notably, whether animals are defined as workers or not, solidarity can still be fostered. Although the concept of solidarity is often used by labor advocates, it is not exclusively applicable to working contexts, nor do both parties or all involved need to be working. Interspecies solidarity can help create change inside and outside of spaces of work and inspire not only different relationships, but societies that advance social solidarity within and across species.

Writers of the Frankfurt School, including Theodor Adorno and Max Horkheimer, provide some further theoretical fodder for an intersectional and solidaristic politics that includes animals. As Ryan Gunderson (2014, 296) argues, "women, ethnic and racial minorities, and workers [were] theorized as co-sufferers with animals due to parallel and intertwined social processes" by these writers. Indeed, Horkheimer argues that "compassion should take the form of solidarity not only with other humans, but also with animals" (Abromeit 2011, 242).[1] Contemporary scholars debate the degree to which solidarity is an emotional or affective process, or an intellectual and political position. Kathy Rudy (2011, 12) argues that "the affective love that connects us to particular animals…can help motivate us to transform the world for animals, not just give them our charity." Like many feminist scholars, in their analysis of workers' organizing, Karen Brodkin and Cynthia Strathman (2004) eschew an emotion-reason binary and instead argue that the battle is over hearts and minds. My view is also that solidarity involves both, and that "the 'emotional' and the 'rational' are not dichotomous, but [rather] are interrelated, and both are integral. Their boundaries are also porous, particularly with concepts like 'fairness' which cannot tidily be defined as exclusively rational or emotional" (Coulter 2013, 193). Precisely how we theorize and define solidarity vis-à-vis emotions, affect, reason, and other terms is less important for me than whether we actually pursue and encourage solidarity in practice.

Solidarity can be felt and thought, but it cannot remain internalized and individualized. Josephine Donovan (2007, 364) writes, "understanding that an animal is in pain or distress—even empathizing or sympathizing with him [or her]—doesn't ensure, however, that the human will act ethically towards the animal. Thus, the originary emotional empathetic response must be supplemented with a political perspective…that enables the human to analyze the situation critically so as to determine who is responsible for the animal suffering, and how that suffering may best be alleviated." As Sally Scholz (2008, 61) puts it, solidarity encourages "not just personal transformation but social transformation." Similarly, Lori Gruen's (2009) call for an active, cultivated process of "empathetic engagement" also points to the need for political as well as intellectual action. Part of what makes the concept of solidarity meaningful and promising is that it is both personal and political, that it is motivational and affirming, and that it can be pursued by one and by many (Coulter 2012; Mallory 2009). Akin to care ethics, interspecies solidarity can be understood as both an activity and a political value (Tronto 2015). Individual acts of solidarity matter, and they can disrupt dominant perceptions and power relations. They can also set a domino effect in motion which propels a broader set of processes. Moreover, solidarity can prompt and inform larger, collective forms of political work. Caring can be and can become political (Briskin 2013; Herd and Meyer 2002; Tronto 1999). I posit that the concept and processes of care work hold great potential for building interspecies solidarity. As noted, care work can be understood and pursued in an instrumental manner. But care work can also be conceptualized as a springboard for fostering more supportive relationships, labor processes, and political projects, including those promoting interspecies solidarity.

A dog in Athens, Greece, usually known as Loukanikos, gained international fame when he kept appearing alongside protesters in mass anti-austerity demonstrations. He was dubbed the riot dog and/or the protest dog, and dozens of news stories, memes, social media accounts, and works of art have been dedicated to him. Perhaps this global fascination with a dog who seemed to be not only supporting protesters but actively participating himself is reflective of a longing for or even a window to a new kind of intersectional, interspecies politics that sees diverse people and animals seeking justice and fairness side by side. Put concisely, interspecies solidarity can be used to envision and implement better conditions for animals, improve people's work lives, and unite human and animal well-being. Interspecies solidarity is both a promise people should make to animals, and an approach to humane animal work with significant promise, one that offers fertile soil, seeds, water, and a vision of what could be cultivated.

Interspecies Solidarity and Praxis

The idea of interspecies solidarity can be put into practice in different ways; it is not a monolithic blueprint to be singularly imposed on all working lives or political projects. Rather, it is an invitation to broaden how labor as both a daily process and a political relationship is understood and approached. Accordingly, interspecies solidarity is both a path and the outline of a destination that encourages new ways of thinking and acting, individually and collectively, that are informed by empathy, support, dignity, and respect. Its precise meaning and applicability will vary across time and space, and be shaped by the particular participants and contexts. In fact, in some communities, it or variants thereof already exist. For example, Elizabeth Sumida Huamana and Laura Alicia Valdiviezob (2014, 79) argue that among some indigenous communities in Peru, "the *gañan*, the farmer who drives the bulls, the *yunta*, who cultivate the earth prior to planting, can be heard encouraging his animals— 'We are going to sweat together.' The philosophy behind this is a sense of interdependence that acknowledges the participation of many healthy elements for a successful crop. As a result, animals are treated kindly, and empathy is strongly cultivated."

Ideas of multispecies mutual dependency and respect are rooted in a number of indigenous cultures and communities; in some cases they endure, in others they have been forced away or below, or become distorted by colonial projects and the imposition of capitalist economics. In countries like Bolivia, the indigenous idea and practice of "living well," of seeing individual beings as interwoven with the health and strength of others and of nature, is being integrated into local and national political projects emphasizing solidarity, complementarity, and reciprocity (Geddes 2014; Löwy 2014; Postero 2013). Such efforts are imperfect, uneven, and complicated by many dynamics, but still underscored by a remarkable challenge to dominant orthodoxies, and a commitment to transforming individuals and societies. Consequently, integral to a process of interspecies solidarity is recognition of these indigenous histories, approaches, and leaders, what lessons they offer and wish to share, as well as how natural-cultural and human-animal relations are also actively being contested, debated, adapted, and remade in indigenous communities today (Beckford, Jacobs, Williams, and Nahdee 2010; McHugh 2013; Powell 2014; Robinson 2010, 2014; Robinson and Wallington 2012).

Moreover, it is not only in indigenous communities where the interconnectedness of human, animal, and environmental concerns have been understood and nurtured. There are nonindigenous farmers and rural communities committed to different kinds of agricultural visions and practices of multispecies respect, and these efforts also offer lessons about alternate paths. Scandinavian and Nordic sociopolitical models reveal the potential of social solidarity to improve people's work lives in a full sense (see, e.g., Lister 2009; Sandberg 2013), and these efforts could be further expanded across species lines. Feminist political economists call for a movement toward caregiving societies, and this could and should include nonhumans (see, e.g., Cohen and Pulkingham 2009; Glenn 2000, 2010; Robinson 2006; Tronto 2013). Nancy Fraser's (1997) specific emphases on the importance of anti-poverty, anti-marginalization, and anti-exploitation politics, respect, and leisure time, could also be somewhat adjusted and thoughtfully expanded to include animals in compelling ways. There are possibilities to be created by building on existing efforts, interweaving currently parallel threads, and imagining new political tapestries.

Without question, the employment of interspecies solidarity challenges us to understand what animals are thinking and feeling, and to change "business as usual" so as to respect them. In fact, the more we see examples of animals transcending their biologically prescribed predatory-prey relationships to forge meaningful relationships with individuals from other species, the more we should act with humility, respect their multifaceted existences, and learn how we, too, can cross alleged divides both within and across species. Sue Donaldson and Will Kymlicka offer these words:

> We can have physically proximate and socially meaningful cooperative relations with animals while still protecting their basic rights. The challenge to developing non-exploitative cooperative relationships is most acutely posed by the case of domesticated animals who are significantly dependent on humans for basic care...[and] we [have] challenged the idea that domesticated animals, by virtue of this dependency on humans, are inherently demeaned, inauthentic,

undignified, oppressed, or unacceptably vulnerable. We argued that dependency per se is not the issue (we are all, after all, dependent and interdependent in complex ways). The issue is how we respond to dependency, individually and as a society... [People] must foster the circumstances and trusting relationships within which animals can exercise agency, and then interpret the signals that animals give regarding their subjective good, preferences, or choices. (Donaldson and Kymlicka 2013, 2–4)

This argument shares the spirit of interspecies solidarity; we are urged to both think and act differently on a daily basis and to organize work differently. In the pursuit of interspecies solidarity, the normalized must be disrupted. Thus, enlisting interspecies solidarity means asking some tough questions about daily work and systems of labor, and its extension will mean certain practices and some whole kinds of animal work cannot be rationalized or sustained. We simply cannot justify requiring a number of species—and individual animals—to work, even if people garner material and/or symbolic gain. In other cases, animals' work may be appropriate, and mutually beneficial, provided that both protections and positive entitlements are afforded (see also Weisberg n.d.). Undoubtedly, the place of killing and death is a particularly complex and significant issue for a range of individuals and communities. My view, overall, is that one cannot kill or condone the killing of someone with whom you feel solidarity, except in cases of self-preservation or mercy. We should support and create healthy, sustainable, and economically sustaining alternatives to labor that mandates killing. Our jobs, industries, and labor relations are socially constructed; they were created by people and can/will be sustained or changed based on our intellectual powers, ethical principles, socioeconomic commitments, and political choices. What is unequivocally clear is that the promise of interspecies solidarity means that animals cannot be seen as subordinates or as tools, and their needs and desires must be taken seriously through changes in perceptions and practices, and through regulation and enforcement. To enlist a concept from labor and feminist struggles, animals also deserve their species' equivalent of bread and roses. Animals want to live. They also want to be happy. Animals have minds, bodies, feelings, desires, and relationships that are connected to and affected by, and simultaneously distinct from their labor. This means that we must not only consider work, but also work-lives and lives, period.

More research and more reflection on the part of those actively involved in animal work spaces will help generate the specifics that are needed for different contexts. Animal welfare researchers already study the lives of working animals, especially equids in the global south (see, e.g., Wade 2014), and the concept of interspecies solidarity could help strengthen, expand, and, in some cases, change their efforts and approaches, or parts thereof. Frontline organizations like The Brooke and some welfare researchers recognize not only equids' physical health but their mental well-being (see, e.g., Geiger and Hovorka 2014). The Five Freedoms that inform much animal welfare research are not exclusively about animals' physical states, as outlined in chapter 3 (see also Boissy et al. 2007; Edgar et al. 2013). Many equid welfare workers in the global south take stock of the current state of animals' psychological health, and also engage local people in understanding work tasks and daily experiences from the perspectives of the animals through frameworks like the Donkey Feeling Analysis and If I Were a Horse exercise, in order to nurture positive practices and interspecies relationships (van Dijk, Pradhan, and Ali 2013). Social workers and some researchers involved in fields like animal-assisted therapy have also been actively analyzing the role of animals in their professional practice, and, notably, considering animals' wellbeing therein (see, e.g., Evans and Gray 2011; Hanrahan 2013; Matsuoka and Sorenson 2013; Serpell, Coppinger, and Fine 2010; Ryan 2011, 2014). Other groups of workers should engage in this kind of reflection for their own fields. Both those who work directly with animals, and those further away such as educators at all levels, policy makers, retailers, chefs, administrators, and workers in agriculture, health care, and social service fields can reflect on both the form and substance of their work and the decisions they make, as part of thinking through ways to eliminate problematic practices, improve human-animal relationships, and envision new ways of building solidaristic multispecies communities. Moreover, those involved in different kinds of care work, those engaged in comparable forms of labor just with different species, and those experiencing shared emotional harm from the daily labor or advocacy work they do, could undoubtedly find much common ground upon which to build. All people should reflect on their consumption choices and what these reject and reward, as well.

The idea of interspecies solidarity can enrich ongoing work, and it also offers a complementary yet expanded way of conceptualizing political work and social change. The advocacy work outlined in chapter 3 is significant because it introduces not only what has been achieved, but how it has been achieved. At the same time, it also reveals how much of the advocacy work being done with/for animals and with/for people who work with animals proceeds without intersecting. Advocates for animals engage with policy makers in the public sector in order to shape their views and actions, but animal-focused activism and workers' forms of collective action rarely unite. In contrast to late-nineteenth-century praxis that often interwove the concerns of groups of people and animals and illustrated an intersectional approach to understandings of both oppression and alternatives, worker and animal advocates today usually follow different paths. There are examples of prominent animal advocates, like Henry Spira, who began their political work in the labour movement, and some union members, activists, and leaders care deeply for animals (Singer 1998). Yet often, proponents of animals' well-being and workers' rights operate in antagonistic spheres. As Claire Jean Kim writes (2015, 19), "most social justice struggles mobilize around a single-optic frame of vision. The process of political conflict then generates a zero-sum dynamic…a *posture of mutual avowal*—an explicit dismissal of and denial of connection with the other form of injustice being raised. This posture…is both ethically and politically troubling." The privileging of certain justice-seeking and deserving groups over others also weakens the prospects of securing meaningful changes, by reproducing fractures and divisions, and thereby ensuring that those with privilege and power who cause oppression can operate comfortably and without substantial, coordinated, forceful opposition.

In spaces of animal work, this mutual avowal stems from both material divisions and conceptual differences, including fundamental differences in what it means to "love" animals. For example, some animal rights activists say they love horses thus condemn horse riding and even equine-assisted therapy if it involves riding. Yet those who work with equines predominantly do so because they "love" horses. Many of those who work with horses find some of the generalizations and claims of outsiders to be ill-informed or ignorant, particularly the suggestion that most contemporary horse-human relationships are about "domination" and "breaking" horses, especially in situations of therapy. These kinds of divergent understandings are not going to vanish, and there will be fundamental contradictions in the worldviews of certain groups of people. However, all involved would benefit from listening and learning more, from having evidence and being accurately informed, and from paying attention to both people and animals' needs.

The concept of interspecies solidarity can be enlisted by advocacy groups, labor unions, and advocates in the public sector. Most importantly, it can help form strengthened ways of thinking about and doing intersecting, mutually supporting political work (Kim 2015). More advocacy groups could think about how human workers are affected by the practices that harm animals. Without question, labor unions could and should expand their ideas and campaigns beyond humans, and I have proposed a number of ways they can begin doing so through union practices, collective bargaining, coalition building, and other forms of political work (Coulter 2014c). As outlined in chapter 3, unions vary in their political orientations and strategies. But a growing number also pursue what is called social unionism by encouraging an expanded vision of who belongs to their community, and by emphasizing a broader politic that includes but also extends beyond workplace-focused and explicitly "economic" issues (Ross 2012). Social unionism is an excellent framework through which to envision and form a larger community of labor that includes animals. Of course, a commitment to interspecies solidarity is not going to immediately reconcile the conceptual and structural contradictions and tensions embedded in current political and economic relations, but it can be used interpersonally and politically as part of forging better ways forward.

Undoubtedly, the ability of workers to advocate for themselves and for animals is context-specific and affected by a number of factors, including the work being done. People working with and reliant on clearly identifiable examples of animals' labor are, in fact, dependent on the animals. Dynamics of partnership and collaboration could be more easily fostered through an interspecies solidarity ethic. In contrast, particularly when the complexities of food, death, and questions about whether animals really can be seen as "working" are most salient, the opportunities for interspecies empathy and connectivity are more constrained and challenging. A slaughterhouse workers' union, for example, is not likely to argue against the killing of animals for food, fur, and leather, for example, but rather to do what it can to improve people's conditions in those workplaces. However, even in the cases where there are union wages and benefits, it is tough to argue that slaughterhouse jobs, particularly those responsible for routinized killing, are or could ever really be good jobs. These are lousy jobs made only moderately more tolerable if better wages are paid. But the work itself cannot improve; it is industrialized killing, dismembering, and packaging. Along with the nonunion positions on factory farms, so much animal agriculture is now a corporatized "commodity chain" which primarily offers difficult, dirty, dangerous work, produces riskier "products," contributes substantial greenhouse gases to the global atmosphere, endangers and often actually pollutes ground water, and keeps billions of animals in situations of normalized suffering and gross indignity. The intensive, for-profit, factory farming system itself exemplifies a kind of banality of evil, to borrow from Hannah Arendt, writ large and multispecies. It is bad for people, animals, the environment, and the health of all three. All deserve better.

The animal politics of unions are not simple. Unions are fought for and formed in particular moments and places, to respond to situations of unfairness and exploitation, and/or to proactively and preemptively self-advocate. Today, labor advocates vary a great deal in their openness to new ideas and strategies, and unions are not homogeneous. As revealed by certain early Teamsters, some unionized workers have opted to use their collective power to try and improve the lives of the animals with whom they worked. Similarly, Porcher (2014) notes that mine horses were seen as coworkers in France, and in 1936 were afforded the right to a week's holiday (at pasture), and the right to retire, at the same time as human workers won key rights. Again, feminist political economy, including the ideas of recognition and redistribution, and the practical history of labor politics are instructive. Many of the rights and regulations in place to protect human workers offer clear guidance about what legal infrastructure could be erected to both protect and benefit animal workers. There is a small but growing number of animal lawyers—animal labor law and lawyers would be useful. We can look at what is afforded to human workers and determine if it ought to be extended to animals. This should include safety standards, breaks, days off, vacations, the right to refuse work on specific days or on a permanent basis, and life after work, among other dimensions. Crucially, in some instances, what we should redistribute is time, self-determination, and the right not to work for people at all. For many animals, the chance to engage in self-initiated and controlled social reproductive and caregiving labor for their own offspring and fellow animals is what is owed. Notably, we can also ask if there are rights, protections, and benefits not yet afforded to either people or animals, but which should be. A positive, progressive labor agenda that fosters interspecies solidarity ought to be about improving the work, work-lives, and lives of people and of animals.

Political and legal changes will only be possible if we think and work differently. And I am by no means the first workers' advocate to point out the intersections of animal and human well-being or to suggest that empathy ought to be extended to nature. Certain activists have pursued direct actions explicitly saying they are doing so in solidarity with animals, and there are advocacy groups that link human and animal concerns, including dignity for poor people and animals, such as Hamilton People and Animal Welfare Solutions. The late Venezuelan president Hugo Chavez explicitly called for "solidarity with animals," and antiracist scholar Angela Davis regularly points out the entanglements of class, race, gender, and species oppression (Hochschartner 2014; Pearson 2014; Vegans of Color 2012). Farm workers' organizer and leader (and vegetarian) Cesar Chavez explicitly confronted agricultural labor and animals' rights:

> We need, in a special way, to work twice as hard to make all people understand that animals are fellow creatures, that we must protect them and love them as we love ourselves. And that's the basis for peace. The basis for peace is respecting all creatures. We cannot hope to have peace until we respect everyone—respect ourselves and respect animals and all living things. We know we cannot defend and be kind to animals until we stop exploiting them—exploiting

them in the name of science, exploiting animals in the name of sport, exploiting animals in the name of fashion, and yes, exploiting animals in the name of food. (Chavez 1992)

Arguments that labor advocates cannot take animals' well-being seriously because jobs are implicated are insufficient and flawed, both politically and ethically. Unionized workers create armaments, from munitions to weaponry, but this does not prevent labor advocates from speaking out and organizing against war. Slogans such as "food not bombs" and "homes not bombs" reflect a long history of critical thinkers and advocates envisioning societies that do not trap working people into defending violence in the name of jobs. Workers who become whistleblowers, from farmers to slaughterhouse workers to aquarium trainers and staff, those who leave paid positions because they can no longer morally justify what they were required to do, also illustrate the potential for refusing to defend injustice in the name of jobs. Such approaches illustrate what Claire Jean Kim (2015) calls multioptic vision, the refusal to privilege one oppression over others, or to fall into the trap of a zero-sum game. A commitment to improving the lives of one group does not mean others are ignored or subordinated. We do not need to defend violence against animals in the name of jobs. In contrast, we must recognize the interconnections and foster expanded, stronger visions and projects of fairness and justice. As Jason Hribal (2007, 110) suggests, "the combination of animals, agency, and class can be a significant and powerful force in the creation of social change," and we should add gender, ethnicity, and race to that mix, at minimum, as well. Workers can challenge the true causes of their hardships and not only seek a bigger piece of the pie, but they can also bake a different and better pie, one made without intra- or interspecies violence that multiple species can enjoy.

Moreover, the inextricability of not only human and animal but environmental concerns is irrefutable. There is alarming epidemiological and public health research that highlights how many contemporary practices in sectors like the live animal trade and industrialized agriculture hurt animals, endanger people's health, and harm the environment. The risks apply to the workers directly involved and the broader public. Central issues include water, air, and soil pollution, increased greenhouse gas production, zoonoses, and antibiotic and microbial resistance (Akhtar 2012; Cutler, Fooks, and Van Der Poel 2010; Blokhuis, Keeling, Gavinelli, and Serratosa 2007; Landers, Cohen, Wittum, and Larson 2012; World Health Organization 2010). As noted, factory farming is a major contributor of climate-change propelling greenhouse gases (Caro, Davis, Bastianoni, and Caldeira 2014; Gerber et al. 2013; Lin et al. 2011; Steinfeld et al. 2006). The issues and lives at stake here are not frivolous or peripheral, nor are they secondary to human workers' well-being; they are inextricably connected to all work and life on this planet. Although animals are the most damaged in such systems, among humans, it is working-class communities and people, women, racialized workers, indigenous peoples, and poor people who are disproportionately and negatively affected. And as climate change worsens and its effects deepen and expand, it is these very people who will continue to be most harmed. Anthropocentric labor researchers and advocates who ignore the intersections of human-animal-environmental issues or who fail to take them seriously are, in fact, abdicating their responsibility to working-class and poor people in a significant and lasting way. Moreover, it is an unjustifiable contradiction to be outraged at corporate greed, worker exploitation, and discrimination against people, and yet to simultaneously condone or ignore the systematic, industrialized decimation of other sentient beings and our shared environment. Those who fight for the marginalized ought to recognize that animals are one of the most oppressed social groups on the planet. Genuine human and social progress and betterment cannot be based on the suffering of others, period. A just and caring society cannot be created on a mass, unmarked animal graveyard.

Given this broader sociopolitical and environmental context, not surprisingly, the critical/Critical animal studies literature is replete with critiques of how people and animals are damaged in for-profit, industrial systems (see, e.g., Adams 2010; Adams and Gruen 2014a, 2014b; Best 2009; Nibert 2013; Nocella, Sorenson, Socha, and Matsuoka 2014; Sorenson 2014; Taylor and Twine 2014; Torres 2007). Larger arguments for a transformed social and economic system are espoused, yet little is said about or proposed for the more immediate or longer term future of work. Health care researchers and practitioners (particularly doctors and veterinarians) have developed an approach called the One Health Model, which interconnects human, animal, and environmental health (see, e.g., Mackenzie, Jeggo, Daszak, and Richt 2013; Woldehanna and Zimicki 2014; Rock and Degeling 2015). This approach can offer conceptual and practical lessons for worlds of work. There is also a clear and irrefutable link between broader questions of fairness and equity, and the prospects for improving animal work spaces. Those interested in combating harm against animals also need to take workplace, intramovement, and societal discrimination based on race, gender, citizenship status, and other factors, seriously (Harper 2009), and recognize the role of economic oppression in perpetuating both people's and animals' suffering. For example, a fruit or vegetable farm does not harm animals, but if the farm workers, including migrant workers, are exploited and denied protections, that is not acceptable.

Therefore, an essential extension of interspecies solidarity is what I call humane jobs: jobs that are good for both people and animals. Indeed, in addition to critique, we need solutions and alternatives. Humane jobs that prioritize both material and experiential well-being and that are about helping rather than harming can exist across sectors, including in agriculture and rural spaces. Were we to expand and create humane jobs, such efforts would contribute to moving the labor force away from jobs that are damaging to people, animals, and the planet, and offer positive alternatives. Humane jobs, simply put, are absolutely integral to more just and sustainable societies and economies, and should play a more central role in labor and animal advocacy projects, job creation and community development plans, and in how we think about work-lives. Intellectual work is thus needed to assess, envision, and develop more and new humane jobs. Such work is a powerful and necessary extension of interspecies solidarity and demonstrates genuine respect for both people and animals.

In sum, I have offered fodder for what ought to be an ongoing, multifaceted conversation and plan of action that continue beyond these pages, and well beyond paper, overall. Although this book ends here, my hope is that this is only the first chapter in a meaningful and transformative story of interspecies solidarity. We can and must do better.

Notes

2 The Work Done by Animals: Identifying and Understanding Animals' Work

1. This book has also been published as *Counting for Nothing: What Men Value and What Women are Worth*. Some neo-Marxists and ecological economists have explored and debated questions of value, as well. See, for instance, Paul Burkett, "Nature's 'Free Gifts' and the Ecological Significance of Value," *Capital & Class* 23, no. 2 (1999): 89–110.

2. See Ryan Gunderson, "The First-Generation Frankfurt School on the Animal Question: Foundations for a Normative Sociological Animal Studies," *Sociological Perspectives* 57, no. 3 (2014): 285–300 for a good discussion of the cultural Marxists of the Frankfurt School and their ideas on animals; and "Marx's Comments on Animal Welfare," *Rethinking Marxism* 23(4): 543–8 for discussion of Marx's views on animal welfare organizations.

Anifesto: The Promise of Interspecies Solidarity

1. For good discussions of the Frankfurt School and animals, see also Christina Gerhardt, "Thinking With: Animals in Schopenhauer, Horkheimer, and Adorno." In *Critical Theory and Animal Liberation*, ed. John Sanbonmatsu. Rowman & Littlefield, 2011: 137–146; Zipporah Weisberg, "The Trouble with Posthumanism: Bacteria Are People, Too." In *Critical Animal Studies: Thinking the Unthinkable*, ed. John Sorenson. Toronto: Canadian Scholars' Press, 2014, 93–116.

References

Abromeit, John. *Max Horkheimer and the Foundations of the Frankfurt School.* Cambridge: Cambridge University Press, 2011.

Adams, Carol J. "Action, Engagement, Remembering—All Together Now." In *Speaking Up for Animals: An Anthology of Women's Voices,* edited by Lisa Kemmerer, ix–xvii. Boulder: Paradigm Publishers, 2012

———. *The Sexual Politics of Meat: A Feminist-Vegetarian Critical Theory.* New York: The Continuum International Publishing Group, 2010.

———. "The War on Compassion." In *The Feminist Care Tradition in Animal Ethics,* edited by Josephine Donovan and Carol J. Adams, 21–36. New York: Columbia University Press, 2007.

Adams, Carol J., and Lori Gruen, eds. *Ecofeminism: Feminist Intersections with Other Animals and the Earth.* New York: Bloomsbury, 2014a.

Adams, Carol J., and Lori Gruen. "Groundwork." In *Ecofeminism: Feminist Intersections with Other Animals and the Earth,* edited by Carol J. Adams and Lori Gruen, 7–36. New York: Bloomsbury, 2014b.

Agriculture and Agri-Food Canada, 2011. "Number of Farms by Industry Type." http://www.agr.gc.ca/poultry/nofrms_eng.htm.

Akhtar, Aysha. *Animals and Public Health: Why Treating Animals Better Is Critical to Human Welfare.* New York: Palgrave Macmillan, 2012.

Allen, Colin, and Marc Bekoff. *Species of Mind: The Philosophy and Biology of Cognitive Ethology.* Cambridge: MIT Press, 1999.

Animal Legal Defense Fund, "ALDF Sues University of Wisconsin over 'Maternal Deprivation' Experiments on Baby Monkeys." October 14, 2014, http://aldf.org/press-room/press-releases/aldf-sues-university-of-wisconsin-over- maternal-deprivation-experiments-on-baby-monkeys/.

Anthony, Raymond. "Farming Animals and the Capabilities Approach: Understanding Roles and Responsibilities Through Narrative Ethics." *Society and Animals* 17, no. 3 (2009): 257–278.

Araiza, Lauren. *To March for Others: The Black Freedom Struggle and the United Farm Workers.* Philadelphia: University of Pennsylvania Press, 2013.

Arluke, Arnold, and Clinton R. Sanders. *Regarding Animals*. Philadelphia: Temple University Press, 1996.

Armstrong, Pat, and Hugh Armstrong. "Public and Private: Implications for Care Work." *The Sociological Review* 53, no. s2 (2005): 167–187.

Armstrong, Pat, Hugh Armstrong, and Krista Scott-Dixon. *Critical to Care: The Invisible Women in Health Services*. Toronto: University of Toronto Press, 2008.

Arluke, Arnold. *Brute Force: Animal Police and the Challenge of Cruelty*. West Lafayette: Purdue University Press, 2004.

———. "Managing Emotions in an Animal Shelter." In *Animals and Human Society: Changing Perspectives*, edited by Aubrey Manning and James Serpell, 145–165. New York: Routledge, 1994.

Ascione, Frank R. *International Handbook of Animal Abuse and Cruelty: Theory, Research, and Application*. West Lafayette: Purdue University Press, 2008.

Ascione, Frank R., and Phil Arkow, eds. *Child Abuse, Domestic Violence, and Animal Abuse: Linking the Circles of Compassion for Prevention and Intervention*. Lafayette: Purdue University Press, 2000.

Baines, Donna. "In a Different Way: Social Unionism in the Nonprofit Social Services—An Australian/Canadian Comparison." *Labor Studies Journal* 35, no. 4 (2010): 480–502.

Bakker, Isabella. "Social Reproduction and the Constitution of a Gendered Political Economy." *New Political Economy* 12, no. 4 (2007): 541–556.

Bakker, Isabella, and Rachel Silvey, eds. *Beyond States and Markets: The Challenges of Social Reproduction*. New York: Routledge, 2012.

Balk, Josh. "A Social Justice Issue We Can Sink Our Teeth into: Factory Farming." *Common Dreams*, 2014: http://www.commondreams.org/view/2014/07/09-2.

Bartram, David J., Ghasem Yadegarfar, and David S. Baldwin. "A Cross-Sectional Study of Mental Health and Well-being and Their Associations in the UK Veterinary Profession." *Social Psychiatry and Psychiatric Epidemiology* 44, no. 12 (2009): 1075–1085.

Bartram, David J., and David S. Baldwin. "Veterinary Surgeons and Suicide: A Structured Review of Possible Influences on Increased Risk." *Veterinary Record* 166, no. 13 (2010): 388–397.

Baur, Gene. *Farm Sanctuary: Changing Hearts and Minds about Animals and Food*. New York: Simon and Schuster, 2008.

BBC News. "Nottinghamshire Police Dogs to Receive 'Pensions.'" *BBC*, November 4, 2013, http://www.bbc.com/news/uk-england-nottinghamshire-24807719.

Beaulieu, Martin S. *Demographic Changes in Canadian Agriculture*. Ottawa: Statistics Canada, 2014.

Beckford, Clinton L., Clint Jacobs, Naomi Williams, and Russell Nahdee. "Aboriginal Environmental Wisdom, Stewardship, and Sustainability: Lessons from the Walpole Island First Nations, Ontario, Canada." *The Journal of Environmental Education* 41, no. 4 (2010): 239–248.

Beers, Diane L. *For the Prevention of Cruelty: The History and Legacy of Animal Rights Activism in the United States.* Athens: Swallow Press/ Ohio University Press, 2006.

Beisser, Andrea L., Scott McClure, Chong Wang, Keith Soring, Rudy Garrison, and Bryce Peckham. "Evaluation of Catastrophic Musculoskeletal Injuries in Thoroughbreds and Quarter Horses at Three Midwestern Racetracks." *Journal of the American Veterinary Medical Association* 239, no. 9 (2011): 1236–1241.

Bekoff, Marc. *The Emotional Lives of Animals: A Leading Scientist Explores Animal Joy, Sorrow, and Empathy-and Why They Matter.* Novato: New World Library, 2008.

Benton, Ted. *Natural Relations: Ecology, Animal Rights and Social Justice.* London: Verso, 1993.

Berget, Bente and Bjarne O. Braastad. "Animal-Assisted Therapy with Farm Animals for Persons with Psychiatric Disorders." *Annali dell'Istituto superiore di Sanità* 47, no. 4 (2011): 384–390.

Berget, Bente, Lena Lidfors, Anna María Pálsdóttir, Katriina Soini, and Karen Thodberg. "Green Care in the Nordic Countries—A Research Field in Progress." Ås: Norwegian University of Life Sciences, 2012.

Best, Steven. "The Rise of Critical Animal Studies: Putting Theory into Action and Animal Liberation into Higher Education." *Journal for Critical Animal Studies* 7, no. 1 (2009): 9–52.

Best, Steven, and Anthony J. Nocella, eds. *Terrorists or Freedom Fighters? Reflections on the Liberation of Animals.* Brooklyn: Lantern Books, 2004.

Bezanson, Kate. *Gender, the State, and Social Reproduction: Household Insecurity in Neo-liberal Times.* Toronto: University of Toronto Press, 2006.

Birke, Lynda. "Hope of Change: Rethinking Human-Animal Relations?" In *Theorizing Animals: Re-thinking Humanimal Relations,* edited by Nik Taylor and Lynda Birke, xvi–xx. Leiden and Boston: Brill, 2011.

———. "Naming Names—Or, What's in It for the Animals?" *Humanimalia* 1, no. 1 (2009): n.p.

Birke, Lynda, Arnold Arluke, and Mike Michael. *The Sacrifice: How Scientific Experiments Transform Animals and People.* West Lafayette: Purdue University Press, 2007.

Bisgould, Lesli, Wendy King, and Jennifer Stopford. "Anything Goes: An Overview of Canada's Legal Approach to Animals on Factory Farms." April, 2001.

Boissy, Alain, Gerhard Manteuffel, Margit Bak Jensen, Randi Oppermann Moe, Berry Spruijt, Linda J. Keeling, and Christoph Winckler. "Assessment of Positive Emotions in Animals to Improve Their Welfare." *Physiology & Behavior* 92, no. 3 (2007): 375–397.

Boris, Eileen, and Rhacel Salazar Parreñas, eds. *Intimate Labors: Cultures, Technologies, and the Politics of Care*. Redwood City: Stanford University Press, 2010.

Bradshaw, G. A., J. G. Borchers, and V. Muller-Paisner. 2012. *Caring for the Caregiver: Analysis and Assessment of Animal Care Professional and Organizational Wellbeing*. Jacksonville: The Kerulos Center.

Brightman, Robert. *Rock Cree Human-Animal Relationships*. Berkeley: University of California Press, 1993.

Briskin, Linda. "In the Public Interest: Nurses on Strike." In *Public Sector Unions in the Age of Austerity*, edited by Stephanie Ross and Larry Savage, 91–102. Halifax and Winnipeg: Fernwood Publishing, 2013.

Briskin, Linda, and Patricia McDermott, eds. *Women Challenging Unions: Feminism, Democracy, and Militancy*. Toronto: University of Toronto Press, 1993.

Broadway, Michael. "Planning for Change in Small Towns or Trying to Avoid the Slaughterhouse Blues." *Journal of Rural Studies* 16, no. 1 (2000): 37–46.

Brodie, Janine. *Politics on the Margins: Restructuring and the Canadian Women's Movement*. Halifax: Fernwood Publishing, 1995.

Brodkin, Karen, and Cynthia Strathmann. "The Struggle for Hearts and Minds: Organization, Ideology, and Emotion." *Labor Studies Journal* 29, no. 3 (2004): 1–24.

The Brooke. "Invisible Helpers: Women's Views on the Contributions of Working Horses, Donkeys, and Mules to Their Lives." London: The Brooke, 2014.

———. "Working Towards Better Wound Management and Feeding" https://www.thebrooke.org/our-work/our-projects2/ethiopias_stone_donkeys.

Brooke-Holmes, Georgina, and Kate Calamatta. "Groom and Employer Survey 2014." Malmesbury: British Grooms Association.

Brosnan, Sarah F., and Frans B. M. De Waal. "Monkeys Reject Unequal Pay." *Nature* 425, no. 6955 (2003): 297–299.

Brueggemann, John, and Cliff Brown. "The Decline of Industrial Unionism in the Meatpacking Industry: Event-Structure Analyses of Labor Unrest, 1946–1987." *Work and Occupations* 30, no. 3 (2003): 327–360.

Budiansky, Stephen. *The Covenant of the Wild: Why Animals Chose Domestication*. New Haven: Yale University Press, 1999.

Buller, Henry. "Individuation, the Mass and Farm Animals." *Theory, Culture & Society* 30, no. 7–8 (2013): 155–175.

Bulliet, Richard W. *Hunters, Herders, and Hamburgers: The Past and Future of Human-Animal Relationships.* New York: Columbia University Press, 2005.

Bunderson, J. Stuart and Jeffery A. Thompson. "The Call of the Wild: Zookeepers, Callings, and the Double-Edged Sword of Deeply Meaningful Work." *Administrative Science Quarterly* 54, no. 1 (2009): 32–57.

Bureau of Labor Statistics. "Occupational Employment and Wages, May 2014, 51–3023 Slaughterers and Meat Packers." http://www.bls.gov/oes/current/oes513023.htm.

Burgon, Hannah Louise. "'Queen of the World': Experiences of 'At-Risk' Young People Participating in Equine-Assisted Learning/Therapy." *Journal of Social Work Practice* 25, no. 2 (2011): 165–183.

Butler, Deborah. "'Not a Job for 'Girly-Girls'': Horseracing, Gender and Work Identities." *Sport in Society* 16, no. 10 (2013): 1309–1325.

Butler, Deborah, and Nickie Charles. "Exaggerated Femininity and Tortured Masculinity: Embodying Gender in the Horseracing Industry." *The Sociological Review* 60, no. 4 (2012): 676–695.

Camarillo, Martha. *Fletcher Street.* Brooklyn: Powerhouse Books, 2006.

Campbell, Ben. "On 'Loving Your Water Buffalo More Than Your Own Mother': Relationships of Animal and Human Care in Nepal." In *Animals in Person: Cultural Perspectives on Human-Animal Intimacies,* edited by John Knight, 79–100. Oxford: Berg, 2005.

Canadian Federation of Human Societies. "Realities of Farming in Canada." http://cfhs.ca/farm/farming_in_canada/.

Canadian Federation of Humane Societies. "Transportation." http://cfhs.ca/farm/transportation/.

Canadian Food and Inspection Agency. "CFIA Statement on Allegations of Inhumane Treatment of Animals at Alberta Facility." http://www.inspection.gc.ca/food/information-for-consumers/fact-sheets/specific-products-and-risks/meat-and-poultry-products/2014-10-10-statement/eng/1412966817511/1412966818652.

Cantin, Anna, and Sylvie Marshall-Lucette. "Examining the Literature on the Efficacy of Equine Assisted Therapy for People with Mental Health and Behavioural Disorders." *Mental Health and Learning Disabilities Research and Practice* 8, no. 1 (2011): 51–61.

Carlsson, Catharina, Daniel Ranta Nilsson, and Bente Traeen. "Equine Assisted Social Work as a Mean for Authentic Relations Between Clients and Staff." *Human-Animal Interaction Bulletin* 2, no. 1 (2014): 19–38.

Caro, Dario, Steven J. Davis, Simone Bastianoni, and Ken Caldeira. "Global and Regional Trends in Greenhouse Gas Emissions from Livestock." *Climatic Change* 126, no. 1–2 (2014): 203–216.

Cassidy, Rebecca. *Horse People: Thoroughbred Culture in Lexington & Newmarket.* Baltimore: The Johns Hopkins University Press, 2007.

———. *The Sport of Kings: Kinship, Class and Thoroughbred Breeding in Newmarket.* Cambridge: Cambridge University Press, 2002.

Castañeda, Heide, Nolan Kline, and Nathaniel Dickey. "Health Concerns of Migrant Backstretch Workers at Horse Racetracks." *Journal of Health Care for the Poor and Underserved* 21, no. 2 (2010): 489–503.

Cavalieri, Paola, and Peter Singer, eds. *The Great Ape Project: Equality Beyond Humanity.* New York: St. Martin's Press, 1994.

Centre for Urban Ecology and Sustainability. "The Value of Bees as Pollinators." http://www.entomology.umn.edu/cues/pollinators/value.html.

Chamberlin, Edward. *Horse: How the Horse Has Shaped Civilizations.* New York: Bluebridge, 2006.

Chavez, Cesar. Speech given upon receiving the Lifetime Achievement Award from In Defense of Animals, 1992. https://www.youtube.com/watch?v=ZeXVjpaNMpk

Chen, Xiaobei. *Tending the Gardens of Citizenship: Child Saving in Toronto, 1880s-1920s.* Toronto: University of Toronto Press, 2005.

City of Athens. "Stray Animals." https://www.cityofathens.gr/en/stray-animals-0.

Clark, Jonathan L. "Labourers or Lab Tools? Rethinking the Role of Lab Animals in Clinical Trials." In *The Rise of Critical Animal Studies: From the Margins to the Centre,* edited by Nik Taylor and Richard Twine, 139–166. London: Routledge, 2014.

Cleaver, Harry. *Reading Capital Politically.* London: Anti-Theses/AK Press, 2000.

Cohen, Marjorie Griffin, and Jane Pulkingham, eds. *Public Policy for Women: The State, Income Security, and Labour Market Issues.* Toronto: University of Toronto Press, 2009.

Cohen, Rachel Lara, Kate Hardy, Teela Sanders, and Carol Wolkowitz. "The Body/Sex/Work Nexus: A Critical Perspective on Body Work and Sex Work." In *Body/Sex/Work: Intimate, Embodied, and Sexual-ized Labour,* edited by Carol Wolkowitz, Rachel Lara Cohen, Teela Sanders, and Kate Hardy, 3–27. Houndsmills: Palgrave Macmillan, 2013.

Collard, Rosemary-Claire. *Animal Traffic: Making, Remaking and Unmaking Commodities in Global Live Wildlife Trade* (PhD Dissertation, University of British Columbia, 2013).

———. "Putting Animals Back Together, Taking Commodities Apart." *Annals of the Association of American Geographers* 104, no. 1 (2014): 151–165.

Corbey, Raymond, and Annette Lanjouw. *The Politics of Species: Reshaping Our Relationships with Other Animals.* Cambridge: Cambridge University Press, 2013.

Corman, Lauren. "History Repeated: Exploitation of Workers and Animals at XL Foods Inc." Paper presented at the United Association for Labor Education Conference. Toronto, Canada, April 20, 2013.

———. "Impossible Subjects: The Figure of the Animal in Paulo Freire's Pedagogy of the Oppressed." *Canadian Journal of Environmental Education* 16 (2012): 29–45.

———. 2005. Interview with Virgil Butler on Animal Voices, February 15. http://animalvoices.ca/2005/02/15/ex-slaughterhouse-worker-virgil-butler/.

Cornu, Jean-Nicolas, Géraldine Cancel-Tassin, Valérie Ondet, Caroline Girardet, and Olivier Cussenot. "Olfactory Detection of Prostate Cancer by Dogs Sniffing Urine: A Step Forward in Early Diagnosis." *European Urology* 59, no. 2 (2011): 197–201.

Costa, Mariarosa Dalla, and Selma James. *The Power of Women and the Subversion of the Community.* Bristol: Falling Wall Press, 1975.

Coulter, Kendra. "Feeling Resistance: Gender and Emotions in Retail Organizing," *WorkingUSA: The Journal of Labor and Society* 16, June (2013a): 191–206.

———. "Herds and Hierachies: Class, Nature, and the Social Construction of Horses in Equestrian Culture." *Society and Animals* 22, no. 2 (2014a): 135–152.

———. "Horse Power: Gender, Work, and Wealth in Canadian Show Jumping." In *Gender and Equestrian Sports,* edited by Miriam Adelman and Jorge Knijnik, 165–181. Dordrecht: Springer International Publishing, 2013b.

———. "How Unions Can Support Pro-Animal Members, Animals and Interspecies Solidarity." *Rabble.ca,* June 11, 2014c. http://rabble.ca/blogs/bloggers/views-expressed/2014/06/how-unions-can-support-pro-animal-members-animals-and-intersp.

———. *Revolutionizing Retail: Workers, Political Action, and Social Change.* New York: Palgrave Macmillan, 2014b.

———. "Solidarity in Deed: Poor People's Organizations, Unions, and the Politics of Antipoverty Work in Ontario." *Anthropology of Work Review* 33, no. 2 (2012): 101–112.

———. "Unionizing Retail: Lessons from Young Women's Grassroots Organizing in the Greater Toronto Area in the 1990s." *Labour/Le Travail* 67, no. 1 (2011): 77–93.

Crick, Bernard. *George Orwell: A Life*. London: Penguin, 1992.

Cripps, Elizabeth. "Saving the Polar Bear, Saving the World: Can the Capabilities Approach Do Justice to Humans, Animals and Ecosystems?" *Res Publica* 16, no. 1 (2010): 1–22.

Cronin, J. Keri. "'Can't You Talk?' Voice and Visual Culture in Early Animal Welfare Campaigns." *Early Popular Visual Culture* 9, no. 3 (2011): 203–223.

———. "'A Mute Yet Eloquent Protest': Visual Culture and Anti-Vivisection Activism in the Age of Mechanical Reproduction." In *Critical Animal Studies: Thinking the Unthinkable*, edited by John Sorenson, 284–297. Toronto: Canadian Scholars' Press, 2014.

CUPE 1600 Toronto Zoo Workers: http://1600.cupe.ca/www/.

Cutler, S. J., A. R. Fooks, and W. H. Van Der Poel. "Public Health Threat of New, Reemerging, and Neglected Zoonoses in the Industrialized World." *Emerging Infectious Diseases* 16, no. 1 (2010): 1–5.

Dabelko-Schoeny, Holly, Gary Phillips, Emily Darrough, Sarah DeAnna, Marie Jarden, Denise Johnson, and Gwendolen Lorch. "Equine-Assisted Intervention for People with Dementia." *Anthrozoos: A Multidisciplinary Journal of the Interactions of People & Animals* 27, no. 1 (2014): 141–155.

Dao, James. 2011. "After Duty, Dogs Suffer Like Soldiers," *New York Times*, December 1. http://www.nytimes.com/2011/12/02/us/more-military-dogs-show-signs-of- combat-stress.html?pagewanted=all&_r=0.

Dave, Naisargi. "Witness: Humans, Animals, and the Politics of Becoming." *Cultural Anthropology*, 29, no. 3 (2014): 433–456.

Davis, Kara, and Wendy Lee, eds. *Defiant Daughters: 21 Women on Art, Activism, Animals, and the Sexual Politics of Meat*. Brooklyn: Lantern Books, 2013.

Davis, Karen. "Thinking Like a Chicken: Farm Animals and the Feminine Connection." In *Animals and Women: Feminist Theoretical Explorations*, edited by Carol J. Adams and Josephine Donovan, 192–212. Durham: Duke University Press, 1995.

Davison, Ron. "Temporary Foreign Worker Changes Worry Meat-Packers." *The Western Producer*, June 23, 2014. http://www.producer.com/daily/temporary-foreign- worker-changes-worry-meat-packers/.

de la Torre, Mónica Padilla, Elodie F. Briefer, Tom Reader, and Alan G. McElligott. "Acoustic Analysis of Cattle (Bos Taurus) Mother-Offspring Contact Calls from a Source-Filter Theory Perspective." *Applied Animal Behaviour Science* 163, no. 1 (2015): 58–68.

de Waal, Frans B. M., and Malini Suchak. "Prosocial Primates: Selfish and Unselfish Motivations." *Philosophical Transactions of the Royal Society B: Biological Sciences* 365, no. 1553 (2010): 2711–2722.

Deemer, Danielle R., and Linda M. Lobao. "Public Concern with Farm Animal Welfare: Religion, Politics, and Human Disadvantage in the Food Sector." *Rural Sociology* 76, no. 2 (2011): 167–196

DeGue, Sarah. "A Triad of Family Violence: Examining Overlap in the Abuse of Children, Partners, and Pets." In *The Psychology of the Human-Animal Bond*, edited by Christopher Blazina, Güler Boyra, David Shen-Miller, 245–262. New York: Springer, 2011.

DeMello, Margo. *Animals and Society: An Introduction to Human-Animal Studies.* New York: Columbia University Press, 2012.

———, ed. *Speaking for Animals: Animal Autobiographical Writing.* New York: Routledge, 2013.

Descola, Philippe. *Beyond Nature and Culture.* Chicago: University of Chicago Press, 2013. Translated by Janet Lloyd.

Desmarais, Annette Aurélie. *La Vía Campesina: Globalization and the Power of Peasants.* Halifax and London: Fernwood Publishing and Pluto Press, 2007.

Dickens, Peter. *Reconstructing Nature: Alienation, Emancipation and the Division of Labour.* London: Routledge, 1996.

Donaldson, Sue, and Will Kymlicka. "Citizen Canine: Agency for Domesticated Animals." Unpublished paper, 2012.

———. "Rethinking Membership and Participation in an Inclusive Democracy: Cognitive Disability, Children, Animals." In *Disability and Political Theory*, edited by Barbara Arneil and Nancy Hirschmann. Cambridge: Cambridge University Press (forthcoming).

———. *Zoopolis: A Political Theory of Animal Rights.* Oxford: Oxford University Press, 2011.

Donovan, Josephine. "Caring to Dialogue: Feminism and the Treatment of Animals (2006)." In *The Feminist Care Tradition in Animal Ethics*, edited by Josephine Donovan and Carol J. Adams, 360–369. New York: Columbia University Press, 2007.

Donovan, Josephine, and Carol J. Adams, eds. *The Feminist Care Tradition in Animal Ethics.* New York: Columbia University Press, 2007.

Duffy, Rosaleen, and Lorraine Moore. "Neoliberalising Nature? Elephant-Back Tourism in Thailand and Botswana." *Antipode* 42, no. 3 (2010): 742–766.

Edgar, Joanne L., Siobhan M. Mullan, Joy C. Pritchard, Una J. C. McFarlane, and David C. J. Main. "Towards a 'Good Life' for Farm Animals: Development of a Resource Tier Framework to Achieve Positive Welfare for Laying Hens." *Animals* 3, no. 3 (2013): 584–605.

Ehmann, R., E. Boedeker, U. Friedrich, J. Sagert, J. Dippon, G. Friedel, and T. Walles. "Canine Scent Detection in the Diagnosis of Lung Cancer:

Revisiting a Puzzling Phenomenon." *European Respiratory Journal* 39, no. 3 (2012): 669–676.

Einwohner, Rachel L. "Gender, Class, and Social Movement Outcomes: Identity and Effectiveness in Two Animal Rights Campaigns." *Gender & Society* 13, no. 1 (1999): 56–76.

Ellis, Colter. "Boundary Labor and the Production of Emotionless Commodities: The Case of Beef Production." *The Sociological Quarterly* 55, no. 1 (2014): 92–118.

———. "The Symbiotic Ideology: Stewardship, Husbandry, and Dominion in Beef Production." *Rural Sociology* 78, no. 4 (2013): 429–449.

Ellis, Colter, and Leslie Irvine. "Reproducing Dominion: Emotional Apprenticeship in the 4-H Youth Livestock Program." *Society and Animals* 18, no. 1 (2010): 21–39.

England, Paula, Michelle Budig, and Nancy Folbre. "Wages of Virtue: The Relative Pay of Care Work." *Social Problems* 49, no. 4 (2002): 455–473.

England, Paula, and Nancy Folbre. "The Cost of Caring." *The Annals of the American Academy of Political and Social Science* 561, no. 1 (1999): 39–51.

Enstad, Nan. *Ladies of Labor, Girls of Adventure: Working Women, Popular Culture, and Labor Politics at the Turn of the Twentieth Century.* New York: Columbia University Press, 1999.

Evans, Erin. "Constitutional Inclusion of Animal Rights in Germany and Switzerland: How Did Animal Protection Become an Issue of National Importance?" *Society and Animals* 18, no. 3 (2010): 231–250.

Evans, Nikki, and Claire Gray. "The Practice and Ethics of Animal-Assisted Therapy with Children and Young People: Is It Enough That We Don't Eat Our Co-Workers?" *British Journal of Social Work* 42 no. 4 (2011): 1–18.

Farid, Sara. "Pakistan's Beasts of Burden." *Reuters*, May 8, 2014. http://blogs.reuters.com/photographers-blog/2014/05/08/pakistans-beasts-of-burden/.

Farm Animal Welfare Council Press Statement. December 5, 1979. http://webarchive.nationalarchives.gov.uk/20121007104210/http://www.fawc.org.uk/pdf/fivefreedoms1979.pdf.

Fennell, David A. *Tourism and Animal Ethics.* London and New York: Routledge, 2012.

Fenton, Andrew. "Can a Chimp Say 'No?'" *Cambridge Quarterly of Healthcare Ethics* 23, no. 2 (2014): 130–139.

Ferguson, Moira. *Animal Advocacy and Englishwomen, 1780–1900: Patriots, Nation, and Empire.* Ann Arbor: University of Michigan Press, 1998.

Figley, Charles R., ed. *Treating Compassion Fatigue.* New York: Routledge, 2002.

Figley, Charles R. and Robert G. Roop. *Compassion Fatigue in the Animal-Care Community.* Washington, DC: Humane Society Press, 2006.

Fine, Aubrey H., ed. *Handbook on Animal-Assisted Therapy: Theoretical Foundations and Guidelines for Practice.* Amsterdam: Academic Press, 2010.

Finsen, Lawrence and Susan Finsen. *The Animal Rights Movement in America: From Compassion to Respect.* New York: Twayne, 1994.

Fisher, Eileen, and Rebecca Reuber. *The State of Entrepreneurship in Canada.* Ottawa: Statistics Canada, 2010.

Fitzgerald, Amy J. "A Social History of the Slaughterhouse: From Inception to Contemporary Implications." *Human Ecology Review* 17, no. 1 (2010): 58–69.

———. "'They Gave Me a Reason to Live': The Protective Effects of Companion Animals on the Suicidality of Abused Women." *Humanity & Society* 31, no. 4 (2007): 355–378.

Fitzgerald, Amy J., Linda Kalof, and Thomas Dietz. "Slaughterhouses and Increased Crime Rates: An Empirical Analysis of the Spillover from 'The Jungle' Into the Surrounding Community." *Organization & Environment* 22, no. 2 (2009): 158–184.

Fitzgerald, Amy J. and Nik Taylor. "The Cultural Hegemony of Meat and the Animal Industrial Complex." In *The Rise of Critical Animal Studies: From the Margins to the Centre,* edited by Nik Taylor and Richard Twine, 165–182. London: Routledge, 2014.

Folbre, Nancy. "Children as Public Goods." *The American Economic Review* 84, no. 2 (1994): 86–90.

Food and Agriculture Organization of the United Nations. "The State of Food and Agriculture: Livestock in the Balance." 2009. http://www.fao.org/docrep/012/i0680e/i0680e.pdf.

Food and Agriculture Organization of the United Nations and The Brooke. *The Role, Impact and Welfare of Working (Traction and Transport) Animals.* Rome: FAO, 2011.

Forrest, Anne. "The Rise and Fall of National Bargaining in the Canadian Meat-Packing Industry." *Relations industrielles/Industrial Relations* 44, no. 2 (1989): 393–408

Forsberg, Lena, and Ulla Tebelius. "The Riding School as a Site for Gender Identity Construction Among Swedish Teenage Girls." *World Leisure Journal* 53, no. 1 (2011): 42–56.

Francione, Gary Lawrence. *Animals as Persons: Essays on the Abolition of Animal Exploitation.* New York: Columbia University Press, 2008.

Francois, Twyla. *Broken Wings: The Breakdown of Animal Protection in the Transportation and Slaughter of Meat Poultry in Canada.* Canadians for the Ethical Treatment of Food Animals, 2009.

Franzway, Suzanne. "Women Working in a Greedy Institution: Commitment and Emotional Labour in the Union Movement." *Gender, Work & Organization* 7, no. 4 (2000): 258–268.

Franzway, Suzanne, and Mary Margaret Fonow. *Making Feminist Politics: Transnational Alliances Between Women and Labor.* Champaign: University of Illinois Press, 2011.

Fraser, Nancy. "From Redistribution to Recognition? Dilemmas of Justice in a 'Post- Socialist' Age." *New Left Review* I, no. 212 (1995): 68–93.

———. *Justice Interruptus: Critical Reflections on the "Postsocialist" Condition.* New York and London: Routledge, 1997.

Frommer, Stephanie S., and Arnold Arluke. "Loving Them to Death: Blame-Displacing Strategies of Animal Shelter Workers and Surrenderers." *Society and Animals* 7, no. 1 (1999): 1–16.

Gaard, Greta, ed. *Ecofeminism: Women, Animals, Nature.* Philadelphia: Temple University Press, 1993.

———. "Ecofeminism Revisited: Rejecting Essentialism and Re-placing Species in a Material Feminist Environmentalism." *Feminist Formations* 23, no. 2 (2011): 26–53.

Gaarder, Emily. "Where the Boys Aren't: The Predominance of Women in Animal Rights Activism." *Feminist Formations* 23, no. 2 (2011a): 54–76.

———. *Women and the Animal Rights Movement.* New Brunswick: Rutgers University Press, 2011b.

Ganz, Marshall. *Why David Sometimes Wins: Leadership, Organization, and Strategy in the California Farm Worker Movement.* Oxford: Oxford University Press, 2009.

Gaynor, Andrea. "Animal Agendas: Conflict over Productive Animals in Twentieth-Century Australian Cities." *Society & Animals* 15, no. 1 (2007): 29–42.

Geddes, Mike. "The Old Is Dying but the New Is Struggling to Be Born: Hegemonic Contestation in Bolivia." *Critical Policy Studies* 8, no. 2 (2014): 165–182.

Geiger, Martha and Alice Hovorka. "Donkey Positionality and Welfare in Botswana." Paper presented at All Creatures Great and Small: Interspecies Interdisciplinary Community, Davis, USA, November 15, 2014.

Gerber, Pierre J., H. Steinfeld, B. Henderson, A. Mottet, C. Opio, J. Dijkman, A. Falcucci, and G. Tempio. *Tackling Climate Change through Livestock: A Global Assessment of Emissions and Mitigation Opportunities.* Rome: Food and Agriculture Organization of the United Nations (FAO), 2013.

Gerhardt, Christina. "Thinking with Animals in Schopenhauer, Horkheimer, and Adorno." In *Critical Theory and Animal Liberation*, edited by John Sanbonmatsu, 137–146. Lanham: Rowman & Littlefield, 2011.

Gillespie, Kathryn. "Sexualized Violence and the Gendered Commodification of the Animal Body in Pacific Northwest US Dairy Production." *Gender, Place & Culture* 21 no. 10 (2014): 1321–1337.

Glasser, Carol L. "Rational Emotions: Animal Rights Theory, Feminist Critiques and Activist Insight." In *The Psychology of the Human-Animal Bond*, edited by Christopher Blazina, Guler Boyra, and David S. Shen-Miller, 307–319. New York: Springer, 2011.

Glenn, Evelyn Nakano. "Creating a Caring Society." *Contemporary Sociology* 29, no. 1 (2000): 84–94.

———. *Forced to Care: Coercion and Caregiving in America*. Cambridge: Harvard University Press, 2010.

Gould, James L., and Carol Grant Gould. *Animal Architects: Building and the Evolution of Intelligence*. New York: Basic Books, 2007.

Greene, Ann Norton. *Horses at Work: Harnessing Power in Industrial America*. Cambridge: Harvard University Press, 2008.

Greenebaum, Jessica. "'I'm Not an Activist!' Animal Rights vs. Animal Welfare in the Purebred Dog Rescue Movement." *Society and Animals* 17, no. 4 (2009): 289–304.

Gregory, James. *Of Victorians and Vegetarians: The Vegetarian Movement in Nineteenth-Century Britain*. London: Tauris Academic Studies, 2007.

Gruen, Lori. "Attending to Nature: Empathetic Engagement with the More Than Human World." *Ethics & the Environment* 14, no. 2 (2009): 23–38.

———. "Dismantling Oppression: An Analysis of the Connection between Women and Animals." In *Ecofeminism: Women, Animals, Nature*, edited by Greta Gaard, 61–90. Philadelphia: Temple University Press, 1993.

———. "Entangled Empathy: An Alternative Approach to Animal Ethics." In *The Politics of Species: Reshaping Our Relationships with Other Animals*, edited by Raymond Corbey and Annette Lanjouw, 223–231. Cambridge: Cambridge University Press, 2013.

———. *Entangled Empathy: An Alternative Ethic for our Relationships with Animals*. Brooklyn: Lantern Books, 2014.

Gullone, Eleonora. *Animal Cruelty, Antisocial Behaviour, and Aggression: More Than a Link*. New York: Palgrave Macmillan, 2012.

Gunderson, Ryan. 2014. "The First-Generation Frankfurt School on the Animal Question: Foundations for a Normative Sociological Animal Studies." *Sociological Perspectives* 57, no. 3 (2014): 285–300.

Halley, Jean O'Malley. *The Parallel Lives of Women and Cows.* New York: Palgrave Macmillan, 2012.

Hamilton, Lindsay A. "The Magic of Mundane Objects: Culture, Identity and Power in a Country Vet's Practice." *The Sociological Review* 61, no. 2 (2013): 265–284.

———. "Muck and Magic: Cultural Transformations in the World of Farm Animal Veterinary Surgeons." *Ethnography* 8, no. 4 (2007): 485–501.

Hamilton, Lindsay A., and Nik Taylor. *Animals at Work: Identity, Politics and Culture in Work with Animals.* Boston and Leiden: Brill Academic Publishers, 2013.

———. "Ethnography in Evolution: Adapting to the Animal 'Other' in Organizations." *Journal of Organizational Ethnography* 1, no. 1 (2012): 43–51.

Hanrahan, Cassandra. "Social Work and Human Animal Bonds and Benefits in Health Research: A Provincial Study." *Critical Social Work* 14, no. 1 (2013): n.p. http://www1.uwindsor.ca/criticalsocialwork/SWhumananimalbonds.

Hansson, H., and C. J. Lagerkvist. "Defining and Measuring Farmers' Attitudes to Farm Animal Welfare." *Animal Welfare* 23, no. 1 (2014): 47–56.

Hatcher, Jessica. "Gorillas, Guns, and Volcanoes: On Patrol with Virunga's First Female Rangers." *The Guardian*, January 15, 2015. http://www.theguardian.com/global-development/2015/jan/15/gorillas-guns-volcanoes-congo-virunga-park.

Haraway, Donna J. *Primate Visions: Gender, Race, and Nature in the World of Modern Science.* New York: Routledge, 1989.

———. *Simians, Cyborgs, and Women: The Reinvention of Nature.* New York: Routledge, 1991.

———. *When Species Meet.* Minneapolis: University of Minnesota Press, 2008.

Harper, A. Breeze. *Sistah Vegan: Food, Identity, Health, and Society: Black Female Vegans Speak.* Brooklyn: Lantern Books, 2009.

Harrod, Howard L. *The Animals Came Dancing: Native American Sacred Ecology and Animal Kinship.* Tucson: University of Arizona Press, 2000.

Hartmann, Heidi I. "The Unhappy Marriage of Marxism and Feminism: Towards a More Progressive Union." *Capital & Class* 3, no. 2 (1979): 1–33.

Harvey, Jean. "Companion and Assistance Animals: Benefits, Welfare and Relationships." *International Journal of Applied Philosophy* 22, no. 2 (2008): 161–176.

Hazel, Susan J., Tania D. Signal, and Nicola Taylor. "Can Teaching Veterinary and Animal-Science Students About Animal Welfare Affect Their Attitude Toward Animals and Human-Related Empathy?" *Journal of Veterinary Medical Education* 38, no. 1 (2011): 74–83.

Hedenborg, Susanna. "Female Jockeys in Swedish Horse Racing 1890–2000: From Minority to Majority—Complex Causes." *The International Journal of the History of Sport* 24, no. 4 (2007): 501–519.

———. "Unknown Soldiers and Very Pretty Ladies: Challenges to the Social Order of Sports in Post-War Sweden." *Sport in History* 29, no. 4 (2009): 601–622.

Hedenborg, Susanna, and Manon Hedenborg White. "From Glamour to Drudgery—Changing Gender Patterns in the Equine Sector: A Comparative Study of Sweden and Great Britain in the Twentieth Century." In *Gender and Equestrian Sports*, edited by Miriam Adelman and Jorge Knijnik, 15–36. Dordrecht: Springer International Publishing, 2013.

Hediger, Ryan, ed. *Animals and War: Studies of Europe and North America*. Boston: Brill, 2012.

Heller, Chaia. *Food, Farms, and Solidarity: French Farmers Challenge Industrial Agriculture and Genetically Modified Crops*. Durham: Duke University Press, 2013.

Herd, Pamela, and Madonna Harrington Meyer. "Care Work: Invisible Civic Engagement." *Gender & Society* 16, no. 5 (2002): 665–688.

Herzog, Harold A. "Gender Differences in Human–Animal Interactions: A Review." *Anthrozoos: A Multidisciplinary Journal of the Interactions of People & Animals* 20, no. 1 (2007): 7–21.

———. "Human Morality and Animal Research: Confessions and Quandaries." *The American Scholar* 62, no. 3 (1993): 337–349.

———. *Some We Love, Some We Hate, Some We Eat: Why It's So Hard to Think Straight about Animals*. New York: Harper Collins, 2010

Hochschartner, Jon. "Vegan Angela Davis Connects Human and Animal Liberation." Counterpunch, January 24, 2014. http://www.counterpunch.org/2014/01/24/vegan-angela-davis-connects-human-and-animal-liberation/.

Hochschild, Arlie Russell. "Emotion Work, Feeling Rules, and Social Structure." *American Journal of Sociology* 85, no. 3 (1979): 551–575.

———. *The Managed Heart*. Berkeley: University of California Press, 1983.

Horse Capture, George P., and Emil Her Many Horses, eds. *A Song for the Horse Nation: Horses in Native American Culture*. New York: Fulcrum Publishing, 2006.

Hribal, Jason C. "Animals, Agency, and Class: Writing the History of Animals From Below." *Human Ecology* 14, no. 1 (2007): 101–112.

———. "Animals Are Part of the Working Class: A Challenge to Labor History." *Labor History* 44, no. 4 (2003): 435–453.

———. *Fear of the Animal Planet: The Hidden History of Animal Resistance.* Oakland: AK Press, 2010.

Hua, Julietta, and Neel Ahuja. "Chimpanzee Sanctuary: 'Surplus' Life and the Politics of Transspecies Care." *American Quarterly* 65, no. 3 (2013): 619–637

Huaman, Elizabeth Sumida, and Laura Alicia Valdiviezo. "Indigenous Knowledge and Education from the Quechua Community to School: Beyond the Formal/Non-Formal Dichotomy." *International Journal of Qualitative Studies in Education* 27, no. 1 (2014): 65–87.

Humane Society of the United States. 2014. "Farm Animal Statistics: Slaughter Totals." http://www.humanesociety.org/news/resources/research/stats_slaughter_totals.html.

Hurn, Samantha. *Humans and Other Animals: Cross-Cultural Perspectives on Human-Animal Interactions.* London: Pluto Press, 2012.

Ingold, Tim. "The Architect and the Bee: Reflections on the Work of Animals and Men." *Man* 18, no. 1 (1983): 1–20.

———. "From Trust to Domination: An Alternative History of Human-Animal Relations." In *Animals and Human Society: Changing Perspectives*, edited by Aubrey Manning and James Serpell, 1–22. New York: Routledge, 1994.

Ingram, Darcy. "Beastly Measures: Animal Welfare, Civil Society, and State Policy in Victorian Canada." *Journal of Canadian Studies/Revue d'études canadiennes* 47, no. 1 (2014): 221–252.

———. *Wildlife, Conservation, and Conflict in Quebec, 1840–1914.* Vancouver: UBC Press, 2013.

International Labour Organization. *ILO Global Estimate of Forced Labour.* Geneva: International Labour Organization, 2012.

Irvine, Leslie. "Animals as Lifechangers and Lifesavers: Pets in the Redemption Narratives of Homeless People." *Journal of Contemporary Ethnography* 42, no. 1 (2013): 3–30.

———. *My Dog Always Eats First: Homeless People and Their Animals.* Boulder: Lynne Rienner, 2013.

Irvine, Leslie, and Jenny R. Vermilya. "Gender Work in a Feminized Profession: The Case of Veterinary Medicine." *Gender & Society* 24, no. 1 (2010): 56–82.

Jasper, James M., and Dorothy Nelkin. *The Animal Rights Crusade: The Growth of a Moral Protest.* New York: Free Press, 1992.

Jones, Pattrice. *Aftershock: Confronting Trauma in a Violent World: A Guide for Activists and their Allies.* Herndon: Lantern Books, 2007.

Kalleberg, Arne L. *Good Jobs, Bad Jobs: The Rise of Polarized and Precarious Employment Systems in the United States, 1970s–2000s.* New York: Russell Sage Foundation, 2011.

Kailo, Kaarina, ed. *Wo(men) and Bears: The Gifts of Nature, Culture and Gender Revisited.* Toronto: Inanna Publications, 2008.

Kalof, Linda. *Looking at Animals in Human History.* London: Reaktion Books, 2007.

Kamioka, Hiroharu, Shinpei Okada, Kiichiro Tsutani, Hyuntae Park, Hiroyasu Okuizumi, Shuichi Handa, and Takuya Oshio. "Effectiveness of Animal-Assisted Therapy: A Systematic Review of Randomized Controlled Trials." *Complementary Therapies in Medicine* 22, no. 2 (2014): 371–390.

Kang, Miliann. "The Managed Hand: The Commercialization of Bodies and Emotions in Korean Immigrant—Owned Nail Salons." *Gender & Society* 17, no. 6 (2003): 820–839.

Kaushik, S. J. "Animals for Work, Recreation and Sports." *Livestock Production Science* 59, no. 2 (1999): 145–154.

Kean, Hilda. *Animal Rights: Political and Social Change in Britain Since 1800.* London: Reaktion Books, 1998

Kemmerer, Lisa, ed. *Sister Species: Women, Animals, and Social Justice.* Urbana: University of Illinois Press, 2011.

———. *Speaking Up for Animals: An Anthology of Women's Voices.* Boulder: Paradigm Publishers, 2012.

Kemmerer, Lisa, and Anthony J. Nocella, eds. *Call to Compassion: Religious Perspectives on Animal Advocacy.* Brooklyn: Lantern Books, 2011.

Kim, Claire Jean. *Dangerous Crossings: Race, Species, and Nature in a Multicultural Age.* Cambridge: Cambridge University Press, 2015.

Kirksey, E. Eben, and Stefan Helmreich, "The Emergence of Multispecies Ethnography." *Cultural Anthropology* 25, no. 4 (2010): 545–576.

Kowalczyk, Agnieszka. "Mapping Non-Human Resistance in the Age of Biocapital." In *The Rise of Critical Animal Studies: From the Margins to the Centre,* edited by Nik Taylor and Richard Twine, 183–200. London: Routledge, 2014.

Knight, John, ed. *Animals in Person: Cultural Perspectives on Human-Animal Intimacies.* Oxford: Berg, 2005.

Labrecque, Jennifer, and Christine A. Walsh. "Homeless Women's Voices on Incorporating Companion Animals into Shelter Services." *Anthrozoos: A Multidisciplinary Journal of The Interactions of People & Animals* 24, no. 1 (2011): 79–95.

Laks, Ellie. *My Gentle Barn: Creating a Sanctuary Where Animals Heal and Children Learn to Hope.* New York: Harmony Books, 2014.

Landers, T. F., B. Cohen, T. E. Wittum, and E. L. Larson. "A Review of Antibiotic Use in Food Animals: Perspective, Policy, and Potential." *Public Health Reports 127*, no. 1 (2012): 4.

Lansbury, Coral. *The Old Brown Dog: Women, Workers, and Vivisection in Edwardian England.* Madison: University of Wisconsin Press, 1985.

Larsen, Elizabeth A. "The Impact of Occupational Sex Segregation on Family Businesses: The Case of American Harness Racing." *Gender, Work & Organization* 13, no. 4 (2006a): 359–382.

———. "A Vicious Oval: Why Women Seldom Reach the Top in American Harness Racing." *Journal of Contemporary Ethnography* 35, no. 2 (2006b): 119–147.

Larson, Greger, Elinor K. Karlsson, Angela Perri, Matthew T. Webster, Simon Ho, Joris Peters, and Peter W. Stahl et al. "Rethinking Dog Domestication by Integrating Genetics, Archeology, and Biogeography." *Proceedings of the National Academy of Sciences* 109, no. 23 (2012): 8878–8883.

Lawrence, Elizabeth Atwood. *Hoofbeats and Society: Studies of Human-Horse Interactions.* Bloomington: Indiana University Press, 1985.

Lee, Paula Young, ed. *Meat, Modernity, and the Rise of the Slaughterhouse.* Hanover: University of New Hampshire Press, 2008.

Lefebvre, Diane, Claire Diederich, Madeleine Delcourt, and Jean-Marie Giffroy. "The Quality of the Relation between Handler and Military Dogs Influences Efficiency and Welfare of Dogs." *Applied Animal Behaviour Science* 104, no. 1 (2007): 49–60.

Lehmann, Wolfgang. *Choosing to Labour? School-Work Transitions and Social Class.* Montréal and Kingston: McGill-Queen's Press, 2007.

Lem, Michelle, Jason B. Coe, Derek B. Haley, Elizabeth Stone, and William O'Grady. "Effects of Companion Animal Ownership among Canadian Street-involved Youth: A Qualitative Analysis." *Journal of Sociology & Social Welfare* 40, no. 4 (2013): 285–304.

Lerner, Henrik, Bo Algers, Stefan Gunnarsson, and Anders Nordgren. "Stakeholders on Meat Production, Meat Consumption and Mitigation of Climate Change: Sweden as a Case." *Journal of Agricultural and Environmental Ethics* 26, no. 3 (2013): 663–678.

Levenson, Karen. *A Comparison of Canadian and EU Welfare Standards.* Toronto: Animal Alliance of Canada, 2011.

Lin, B. B., M. J. Chappell, J. Vandermeer, G. Smith, E. Quintero, R. Bezner-Kerr, and D. M. Griffith et al. "Effects of Industrial Agriculture on Climate Change and the Mitigation Potential of Small-Scale Agro-Ecological Farms." *CAB Reviews* 6, no. 20 (2011): 1–18.

Linzey, Andrew. *Creatures of the Same God: Explorations in Animal Theology*. Brooklyn: Lantern Books, 2009.

——, ed. *The Link between Animal Abuse and Human Violence*. Eastbourne: Sussex Academic Press, 2009.

Lister, Ruth. "A Nordic Nirvana? Gender, Citizenship, and Social Justice in the Nordic Welfare States." *Social Politics: International Studies in Gender, State & Society* 16, no. 2 (2009): 242–278.

Little, Margaret Hillyard. "Militant Mothers Fight Poverty: The Just Society Movement, 1968–1971." *Labour/Le Travail* 59 (2007): 179–197.

Lochi, Ghulam Murtaza, Muhammad Ghias Uddin Shah, Muhammad Shuaib Khan, Jamil Ahmad Gandahi, Dildaar Husain Kalhorro, Sumera Ali Khan, Farooq Alam, Abdul Manan Khokar, Mubashir Hasan, and Abdul Haseeb Danish. "Management and Welfare Needs of Donkeys in the Rural Areas of Noushahro Feroze, Pakistan." *Scientific Research and Essays* 9, no. 10 (2014): 410–413.

Löwy, Michael. "Ecosocial Struggles of Indigenous Peoples." *Capitalism Nature Socialism* 25 no. 2 (2014): 14–24.

Luxton, Meg. *More Than a Labour of Love: Three Generations of Women's Work in the Home*. Toronto: Women's Press-Canadian Scholars' Press Inc., 2009.

Luxton, Meg, and Kate Bezanson, eds. *Social Reproduction: Feminist Political Economy Challenges Neo-liberalism*. Montréal and Kingston: McGill-Queen's University Press, 2006.

Lyderson, Kari. "A String of Slaughterhouse Success for UFCW." *In These Times*, December 6, 2011. http://inthesetimes.com/working/entry/12372/slaughterhouse_successes_for_ufcw.

Mackenzie, John S., Martyn Jeggo, Peter S. Daszak, and Juergen A. Richt. *One Health: The Human-Animal-Environment Interfaces in Emerging Infectious Diseases*. Berlin: Springer, 2013.

MacLachlan, Ian. *Kill and Chill: Restructuring Canada's Beef Commodity Chain*. Toronto: University of Toronto Press, 2001.

Mallory, Chaone. "Val Plumwood and Ecofeminist Political Solidarity: Standing with the Natural Other." *Ethics & the Environment* 14, no. 2 (2009): 3–21.

Markovits, Andrei S., and Robin Queen. "Women and the World of Dog Rescue: A Case Study of the State of Michigan." *Society and Animals* 17, no. 4 (2009): 325–342.

Markovits, Andrei S. and Katherine Crosby. *From Property to Family: American Dog Rescue and the Discourse of Compassion*. Ann Arbor: University of Michigan Press, 2014.

Marohn, Stephanie. *What the Animals Taught Me*. Charlottesville: Hampton Roads, 2012.

Marx, Karl, and Friedrich Engels. n.d. "Machinery and Modern Industry" in *Capital Volume One*: https://www.marxists.org/archive/marx/works/1867-c1/ch15.htm.

———. *The Marx-Engels Reader*, edited by Robert C. Tucker. New York: W. W. Norton, 1978.

Matsuoka, Atsuko, and John Sorenson. "Human Consequences of Animal Exploitation: Needs for Redefining Social Welfare." *Journal of Sociology & Social Welfare* 40, no. 4 (2013): 7–32.

Maurstad, Anita, Dona Davis, and Sarah Cowles. "Co-being and Intra-action in Horse–Human Relationships: A Multi-species Ethnography of Be(com)ing Human and Be(com)ing Horse." *Social Anthropology* 21, no. 3 (2013): 322–335.

McArthur, Jo-Anne. *We Animals*. Brooklyn: Lantern Books, 2014.

McCabe, Edward R. B. "2009 Presidential Address: Beyond Darwin? Evolution, Coevolution, and the American Society of Human Genetics." *American Journal of Human Genetics* 86, no. 3 (2010): 311–215.

McChesney, Robert W., and Ben Scott. "Introduction" to Sinclair, Upton. *The Brass Check: A Study of American Journalism*. Champaign: University of Illinois Press, 2002.

McFarland, Sarah E., and Ryan Hediger, eds. *Animals and Agency: An Interdisciplinary Exploration*. Leiden: Brill, 2009.

McGrath, S. Siobhán, and Kendra Strauss. "Unfreedom and Workers' Power: Ever-Present Possibilities." In *The International Political Economy of Production* edited by Kees van der Pijl, 299–317. Cheltenham: Edward Elgar, 2015 (forthcoming).

McHugh, Susan. *Animal Stories: Narrating Across Species Lines*. Minneapolis: University of Minnesota Press, 2011.

———. "A Flash Point in Inuit Memories: Endangered Knowledges in the Mountie Sled Dog Massacre." *ESC: English Studies in Canada* 39, no. 1 (2013): 149–175.

M'Closkey, Kathy. *Swept under the Rug: A Hidden History of Navajo Weaving*. Albuquerque: University of New Mexico Press, 2002.

McShane, Clay, and Joel Tarr. *The Horse in the City: Living Machines in the Nineteenth Century*. Baltimore: Johns Hopkins University Press, 2007.

Merskin, Debra. "Hearing Voices: The Promise of Participatory Action Research for Animals." *Action Research* 9, no. 2 (2011): 144–161.

Mgode, Georgies F., Bart J. Weetjens, Thorben Nawrath, Christophe Cox, Maureen Jubitana, Robert S. Machang'u, and Stéphan Cohen-Bacrie et al. "Diagnosis of Tuberculosis by Trained African Giant Pouched Rats and Confounding Impact of Pathogens and Microflora of the Respiratory Tract." *Journal of Clinical Microbiology* 50, no. 2 (2012): 274–280.

Miller, Janet. "Racing Bodies." In *Body/Sex/Work: Intimate, Embodied, and Sexualized Labour*, edited by Carol Wolkowitz, Rachel Lara Cohen, Teela Sanders, and Kate Hardy, 193–206. Houndmills: Palgrave Macmillan, 2013a.

———. "Resistance Is Futile? Evidence from the Small Firms Sector." Paper presented at the *International Labour Process Conference*, New Brunswick, USA, 18–20 March 2013b.

———. "Turf Wars: Stable Lads' Strikes and Union Recognition in the Twentieth Century." *Historical Studies in Industrial Relations* 34 (2013c): 111–140.

———. "'We Can't Join a Union, That Would Harm the Horses': Worker Resistance in the UK Horseracing Industry." *Centre for Employment Studies Reseach Review* (April 2008): 1–4.

Misión Nevado, ¿Que es la Misión Nevado? http://misionnevado.blogspot.ca/p/que-es-la-mision-nevado.html.

Montgomery, Charlotte. *Blood Relations: Animals, Humans, and Politics.* Toronto: Between the Lines, 2000.

Morgan, Lewis Henry. *The American Beaver and His Works.* Philadelphia: J.B. Lipincott & Co., 1868.

Morris, Brian. *The Power of Animals: An Ethnography.* Oxford: Berg, 2000.

Mullin, Molly H. "Mirrors and Windows: Sociocultural Studies of Human-Animal Relationships." *Annual Review of Anthropology* 28 (1999): 201–224.

Murray, Mary. "The Underdog in History: Serfdom, Slavery and Species in the Creation and Development of Capitalism." In *Theorizing Animals: Re-thinking Humanimal Relations*, edited by Nik Taylor and Tania Signal, 87–106. Boston: Brill, 2011.

Nance, Susan. *Entertaining Elephants: Animal Agency and the Business of the American Circus.* Baltimore: Johns Hopkins University Press, 2013a.

———. "Game Stallions and Other 'Horseface Minstrelsies' of the American Turf." *Theatre Journal* 65, no. 3 (2013b): 355–372.

———. "'A Star is Born to Buck': Animal Celebrity and the Marketing of Professional Rodeo." In *Sport, Animals, and Society*, edited by James Gillett and Michelle Gilbert, 173–191. New York: Routledge, 2014.

Naples, Nancy A. *Grassroots Warriors: Activist Mothering, Community Work, and the War on Poverty.* New York: Routledge, 1998.

Neme, Laurel. "For Rangers on the Front Lines of Anti-Poaching War, Daily Trauma." *National Geographic*, June 14, 2014. http://news.nationalgeographic.com/news/2014/06/140627-congo-virunga-wildlife-rangers-elephants-rhinos-poaching/.

Neumann, Sandra L. "Animal Welfare Volunteers: Who Are They and Why Do They Do What They Do?" *Anthrozoos: A Multidisciplinary Journal of The Interactions of People & Animals* 23, no. 4 (2010): 351–364.

Nibert, David A. *Animal Oppression and Human Violence: Domesecration, Capitalism, and Global Conflict.* New York: Columbia University Press, 2013.

———. "Animals, Immigrants, and Profits: Slaughterhouses and the Political Economy of Oppression." In *Critical Animal Studies: Thinking the Unthinkable*, edited by John Sorenson, 3–17. Toronto: Canadian Scholars' Press, 2014.

Niven, Charles D. *History of the Humane Movement.* New York: Transatlantic Press, 1967.

Nocella II, Anthony J., Colin Salter, and Judy K. C. Bentley, eds. *Animals and War: Confronting the Military-Animal Industrial Complex.* Lanham: Lexington Books, 2013.

Nocella, Anthony, John Sorenson, Kim Socha, and Atsuko Matsuoka, eds. *Defining Critical Animal Studies: An Intersectional Social Justice Approach for Liberation.* New York: Peter Lang, 2014.

Noske, Barbara. *Beyond Boundaries: Humans and Animals.* Montréal: Black Rose Books, 1997.

———. *Humans and Other Animals: Beyond the Boundaries of Anthropology.* London: Pluto Press, 1989.

Nussbaum, Martha C. *Frontiers of Justice: Disability, Nationality, Species Membership.* Cambridge: Belknap Press of Harvard University Press, 2006.

Onion, Rebecca. "Sled Dogs of the American North: On Masculinity, Whiteness, and Human Freedom." In *Animals and Agency: An Interdisciplinary Framework*, edited by Sarah E. McFarland and Ryan Hediger, 129–156. Leiden: Brill, 2009.

Pachirat, Timothy. *Every Twelve Seconds: Industrialized Slaughter and the Politics of Sight.* New Haven: Yale University Press, 2011.

Parreñas, Rheana. "Producing Affect: Transnational Volunteerism in a Malaysian Orangutan Rehabilitation Center." *American Ethnologist* 39, no. 4 (2012): 673–687.

Paterniti, Michael. "The Dogs of War." *National Geographic*, August 2014. http://ngm.nationalgeographic.com/2014/06/war-dogs/paterniti-text.

Patterson, Francine, and Eugene Linden. *The Education of Koko.* New York: Holt, Winehart and Winston, 1981.

Pearson, Tamara. "Venezuelan Government Creates Mission for Street Animals." *Venezuela Analysis*, January 7, 2014. http://venezuelanalysis.com/news/10268.

Peek, Charles W., Nancy J. Bell, and Charlotte C. Dunham. "Gender, Gender Ideology, and Animal Rights Advocacy." *Gender & Society* 10, no. 4 (1996): 464–478.

Peggs, Kay. "The 'Animal-Advocacy Agenda': Exploring Sociology for Non-human Animals." *The Sociological Review* 61, no. 3 (2013): 591–606.

———. *Animals and Sociology.* New York: Palgrave Macmillan, 2012.

Pendry, Patricia, Annelise N. Smith, and Stephanie M. Roeter. "Randomized Trial Examines Effects of Equine Facilitated Learning on Adolescents' Basal Cortisol Levels." *Human-Animal Interaction Bulletin* 2, no. 1 (2014): 80–95.

Perlo, Katherine. "Marx and the Underdog." *Society and Animals* 10, no. 3 (2002): 303–318.

Phillips, Michael "Even His Red Squeak Toy Can't Get First Sgt. Gunner, USMC, to Fight." *Wall Street* Journal, March 3, 2010. http://online.wsj.com/news/articles/SB10001424052748704479404575087360790295570.

Pini, Barbara, and Belinda Leach. "Transformations of Class and Gender in the Globalized Countryside: An Introduction." In *Reshaping Gender and Class in Rural Spaces*, edited by Barbara Pini and Belinda Leach, 1–23. Farnham: Ashgate, 2011.

Piven, Frances Fox, and Richard A. Cloward. *Poor People's Movements: Why They Succeed, How They Fail.* New York: Vintage Books, 1979.

Plumwood, Val. *Environmental Culture: The Ecological Crisis of Reason.* London and New York: Routledge, 2002.

Pomedli, Michael. *Living with Animals: Ojibwe Spirit Power.* Toronto: University of Toronto Press, 2014.

Porcher, Jocelyne. "Animal Work." In *The Oxford Handbook of Animal Studies*, edited by Linda Kalof. New York: Oxford University Press, forthcoming 2016.

———. "The Relationship Between Workers and Animals in the Pork Industry: A Shared Suffering." *Journal of Agricultural and Environmental Ethics* 24, no. 1 (2011): 3–17.

———. "The Work of Animals: A Challenge for Social Sciences." *Humanimalia: A Journal of Human-Animal Interface Studies* 6, no. 1 (2014): 1–9.

Porcher, Jocelyne, and Schmitt, Tiphaine. "Dairy Cows: Workers in the Shadows? *Society and Animals* 20, no. 1 (2012): 39–60.

Postero, Nancy. "Protecting Mother Earth in Bolivia: Discourse and Deeds in the Morales Administration." In *Amazonia: Environment and the Law in Amazonia: A Plurilateral Encounter*, edited by James

M. Cooper and Christine Hunefelt, 78–93. Portland: Sussex Academic Press, 2013.

Powell, Dylan. "Veganism in the Occupied Territories: Anti-Colonialism and Animal Liberation." http://dylanxpowell.com/2014/03/01/veganism-in-the-occupied-territories-anti-colonialism-and-animal-liberation/.

Preece, Rod, and Lorna Chamberlain. *Animal Welfare and Human Values.* Waterloo: Wilfrid Laurier University Press, 1993.

Regan, Tom. *The Case for Animal Rights.* Dordrecht: Springer, 1987.

Renold, Emma, and Gabrielle Ivinson. "Horse-Girl Assemblages: Towards a Post-Human Cartography of Girls' Desire in an Ex-Mining Valleys Community." *Discourse: Studies in the Cultural Politics of Education* 35, no. 3 (2014): 361–376.

Ritvo, Harriet. *The Animal Estate: The English and Other Creatures in the Victorian Age.* Cambridge: Harvard University Press, 1987.

———. *Noble Cows and Hybrid Zebras: Essays on Animals and History.* Charlottesville: University of Virginia Press, 2010.

Robinson, Catherine J., and Tabatha J. Wallington. "Boundary Work: Engaging Knowledge Systems in Co-Management of Feral Animals on Indigenous Lands." *Ecology and Society* 17, no. 2 (2012): 16–28.

Robinson, Fiona. "Beyond Labor Rights: The Ethics of Care and Women's Work in the Global Economy." *International Feminist Journal of Politics* 8, no. 3 (2006): 321–342.

Robinson, Margaret. "Animal Personhood in Mi'kmaq Perspective." *Societies* 4, no. 4 (2014): 672–688.

———. "Veganism and Mi'kmaq Legends: Feminist Natives Do Eat Tofu." Paper presented at the American Academy of Religion conference, October, 2010, Atlanta, USA.

Rock, Melanie J., and Chris Degeling. "Public Health Ethics and More-Than-Human Solidarity." *Social Science & Medicine* 129 (2015): 61–67.

Rogelberg, Steven G., Natalie DiGiacomo, Charlie L. Reeve, Christiane Spitzmüller, Olga L. Clark, Lisa Teeter, Alan G. Walker, Nathan T. Carter, and Paula G. Starling. "What Shelters Can Do about Euthanasia-Related Stress: An Examination of Recommendations from Those on the Front Line." *Journal of Applied Animal Welfare Science* 10, no. 4 (2007): 331–347.

Roine, Antti, Erik Veskimäe, Antti Tuokko, Pekka Kumpulainen, Juha Koskimäki, Tuomo A. Keinänen, Merja R. Häkkinen et al. "Detection of Prostate Cancer by an Electronic Nose: A Proof of Principle Study." *The Journal of Urology* 192 no. 1 (2014): 230–235.

Roseberry, William. *Anthropologies and Histories: Essays in Culture, History, and Political Economy.* New Brunswick and London: Rutgers University Press, 1989.

Ross, Stephanie. "Business Unionism and Social Unionism in Theory and Practice." In *Rethinking the Politics of Labour in Canada*, edited by Stephanie Ross and Larry Savage, 33–46. Halifax: Fernwood Press, 2012.

Royal Society for the Prevention of Cruelty to Animals. "Our History." http://www.rspca.org.uk/utilities/aboutus/history.

Rudy, Kathy. *Loving Animals: Toward a New Animal Advocacy*. Minneapolis: University of Minnesota Press, 2011.

Rutman, Andrew, and Leonard Jones. *In the Children's Aid: J.J. Kelso and Child Welfare in Ontario*. Toronto: University of Toronto Press, 1981.

Ryan, Thomas. *Animals and Social Work: A Moral Introduction*. New York: Palgrave Macmillan, 2011.

———. ed. *Animals in Social Work: Why and How They Matter*. New York: Palgrave Macmillan, 2014.

Sandberg, Äke, ed. *Nordic Lights: Work, Management and Welfare in Scandinavia*. Stockholm: SNS förlag, 2013.

Sanders, Clinton R. *Understanding Dogs: Living and Working with Canine Companions*. Philadelphia: Temple University Press, 1999.

———. "Working Out Back: The Veterinary Technician and 'Dirty Work.'" *Journal of Contemporary Ethnography* 39, no. 3 (2010): 243–272.

Sangster, Joan. *Transforming Labour: Women and Work in Post-War Canada*. Toronto: University of Toronto Press, 2010.

Savage-Rumbaugh, Sue, Kanzi Wamba, Panbanisha Wamba, and Nyota Wamba. "Welfare of Apes in Captive Environments: Comments on, and by, a specific Group of Apes." *Journal of Applied Animal Welfare Science* 10, no. 1 (2007): 7–19.

Scanlan, Lawrence. *The Horse That God Built: The Untold Story of Secretariat*. Toronto: HarperCollins, 2006.

Scholz, Sally J. *Political Solidarity*. University Park, PA: Penn State Press, 2008.

Schwartzman, Kathleen C. *The Chicken Trail: Following Workers, Migrants, and Corporations Across the Americas*. Ithaca, NY: Cornell University Press, 2013.

Scott, Shelly R. "The Racehorse as Protagonist: Agency, Independence, and Improvisation." In *Animals and Agency: An Interdisciplinary Framework*, edited by Sarah E. McFarland and Ryan Hediger, 45–66. Leiden: Brill, 2009.

Sempik, J., R. E. Hine, and D. Wilcox. *Green Care: A Conceptual Framework*. Loughborough University, 2010.

Serpell, J. A., R. Coppinger, and A. H. Fine. "Welfare Considerations in Therapy and Assistance Animals." In *Handbook on Animal-Assisted Therapy: Theoretical Foundations and Guidelines for Practice*, edited by Aubrey H. Fine, 481–502. London: Academic Press, 2006.

Shantz, Jeff. *Green Syndicalism: An Alternative Red-Green Vision.* Syracuse: Syracuse University Press, 2012.

Shaw, David Gary. "The Torturer's Horse: Agency and Animals in History." *History and Theory* 52, no. 4 (2013): 146–167.

Shaw, Randy. *Beyond the Fields: Cesar Chavez, the UFW, and the Struggle for Justice in the 21st Century.* Berkeley: University of California Press, 2008.

Shipman, Pat. *Animal Connection: A New Perspective on What Makes Us Human.* New York: Norton, 2011.

———. "Do the Eyes Have It? Dog Domestication May Have Helped Humans Thrive While Neandertals Declined." *American Scientist* 100, no. 3 (2012): 198–205.

Shiva, Vandana. *Stolen Harvest: The Hijacking of the Global Food Supply.* Cambridge: South End Press, 2000.

Shukin, Nicole. *Animal Capital: Rendering Life in Biopolitical Times.* Minneapolis: University of Minnesota Press, 2009.

Silverstein, Helena. *Unleashing Rights: Law, Meaning, and the Animal Rights Movement.* Ann Arbor: University of Michigan Press, 1996.

Singer, Peter. *Animal Liberation: A New Ethics for Our Treatment of Animals.* New York: Avon Books, 1975.

———. *Ethics into Action: Henry Spira and the Animal Rights Movement.* Lanham: Rowman & Littlefield, 1998.

———, ed. *In Defense of Animals: The Second Wave.* Malden: Blackwell, 2006.

Skipper, Gregory E. and Jerome B. Williams. "Failure to Acknowledge High Suicide Risk Among Veterinarians." *Journal of Veterinary Medical Education* 39, no. 1 (2012): 79–82.

Smart, Alan. "Critical Perspectives on Multispecies Ethnography." *Critique of Anthropology* 34, no. 1 (2014): 3–7.

Smith, Bonnie G. *The Gender of History: Men, Women, and Historical Practice.* Cambridge: Harvard University Press, 1998.

Smith, Kimberly K. *Governing Animals: Animal Welfare and the Liberal State.* Oxford: Oxford University Press, 2012.

Smuts, Barbara. "Reflections" in Coetzee, J.M. *The Lives of Animals*, 107–120. Princeton: Princeton University Press, 1999.

Sodikoff, Genese. "The Low-Wage Conservationist: Biodiversity and Perversities of Value in Madagascar." *American Anthropologist* 111, no. 4 (2009): 443–455.

Sorenson, John, ed. *Critical Animal Studies: Thinking the Unthinkable.* Toronto: Canadian Scholars' Press, 2014.

Stallwood, Kim. *Growl: Life Lessons, Hard Truths, and Bold Strategies from an Animal Advocate.* Brooklyn: Lantern Books, 2014.

Statistics Canada. 2011a. "Canada's Farm Population: Agriculture-Population Linkage Data from the 2006 Census." http://www.statcan.gc.ca/ca-ra2006/agpop/article-eng.htm.

———. 2011b. CANSIM, 2011, Table 111–008 and 111–009.

Steinfeld, Henning, Pierre Gerber, Tom Wassenaar, Vincent Castel, Mauricio Rosales, and Cees de Haan. *Livestock's Long Shadow: Environmental Issues and Options*. Rome: Food and Agriculture Organization of the United Nations, 2006.

Stuart, Diana, Rebecca Schewe, and Ryan Gunderson. "Extending Social Theory to Farm Animals: Addressing Alienation in the Dairy Sector." *Sociologia Ruralis* 53, no. 2 (2013): 201–222.

Stuart, Tristram. *The Bloodless Revolution: Radical Vegetarians and the Discovery of India*. London: Harper Press, 2006

Stull, Donald D., and Michael J. Broadway. *Slaughterhouse Blues: The Meat and Poultry Industry in North America*, 2nd edn. Belmont: Wadsworth Cengage Learning, 2013.

Stull, Donald D., Michael J. Broadway, and David Griffith. *Anyway You Cut It: Meatpacking and Small-Town America*. Lawrence: University Press of Kansas, 1995.

Sunstein, Cass R., and Martha C. Nussbaum, eds. *Animal Rights: Current Debates and New Directions*. Oxford: Oxford University Press, 2004.

Swabe, Joanna. *Animals, Disease and Human Society: Human-Animal Relations and the Rise of Veterinary Medicine*. New York: Routledge, 1999.

Swann, William J. "Improving the Welfare of Working Equine Animals in Developing Countries." *Applied Animal Behaviour Science* 100, no. 1 (2006): 148–151.

Tait, Vanessa. *Poor Workers' Unions: Rebuilding Labor From Below*. Cambridge: South End Press, 2005.

Taylor, Nicola. "In it for the Nonhuman Animals: Animal Welfare, Moral Certainty, and Disagreements." *Society and Animals* 12, no. 4 (2004): 317–339.

Taylor, Nik. "Animal Shelter Emotion Management A Case of In Situ Hegemonic Resistance?" *Sociology* 44, no. 1 (2010): 85–101.

———. *Humans, Animals, and Society: An Introduction to Human-Animal Studies*. New York: Lantern Books, 2013.

Taylor, Nik, and Richard Twine. "Introduction: Locating the 'Critical' in Critical Animal Studies." In *The Rise of Critical Animal Studies: From the Margins to the Centre*, edited by Nik Taylor and Richard Twine, 1–15. London: Routledge, 2014.

Teamsters. "Teamsters Visual History Timeline." http://teamster.org/content/teamster-history-visual-timeline.

Theodossopoulos, Dimitrios. "Care, Order, and Usefulness: The Context of the Human- Animal Relationship in a Greek Island Community." In *Animals in Person: Cultural Perspectives on Human-Animal Intimacies*, edited by John Knight, 15–36. Oxford: Berg, 2005.

Thompson, Kirilly, and Lynda Birke. "The Horse Has Got to Want to Help: Human-Animal Habituses and Networks in Amateur Show Jumping." In *Sport, Animals, and Society*, edited by James Gillett and Michelle Gilbert, 69–84. New York: Routledge, 2014.

Tiplady, C. M., D. B. Walsh, and C. J. C. Phillips. "Intimate Partner Violence and Companion Animal Welfare." *Australian Veterinary Journal* 90, no. 12 (2012): 48–53.

Torres, Bob. *Making a Killing: The Political Economy of Animal Rights*. Oakland: AK Press, 2007.

Tronto, Joan C. *Caring Democracy: Markets, Equality, and Justice*. New York: NYU Press, 2013.

———. "Care Ethics." *The Encyclopedia of Political Thought*. Wiley Online Library, forthcoming 2015.

———. *Moral Boundaries: A Political Argument for an Ethic of Care*. London: Psychology Press, 1993.

Twine, Richard. "Animals on Drugs: Understanding the Role of Pharmaceutical Companies in the Animal-Industrial Complex." *Journal of Bioethical Inquiry* 10, no. 4 (2013): 505–514.

———. "Revealing the 'Animal-Industrial Complex': A Concept and Method for Critical Animal Studies." *Journal for Critical Animal Studies* 10, no. 1 (2012): 12–39.

van Dijk, Lisa, S. K. Pradhan, Murad Ali, and Ramesh Ranjan. "Sustainable Animal Welfare: Community-led Action for Improving Care and Livelihoods." In *Participatory Learning and Action: Tools for Supporting Sustainable Natural Resource Management and Livelihoods*, edited by Holly Ashley, Nicole Kenton, and Angela Milligan, 37–50. London: International Institute for Environment and Development, 2013.

van Schaik, Carel P., Marc Ancrenaz, Gwendolyn Borgen, Birute Galdikas, Cheryl D. Knott, Ian Singleton, Akira Suzuki, Sri Suci Utami, and Michelle Merrill. "Orangutan Cultures and the Evolution of Material Culture." *Science* 299, no. 5603 (2003): 102–105.

Vegans of Color. "Angela Davis on Eating Chickens, Occupy, and Including Animals in Social Justice Initiatives of the 99%." March 3, 2012. https://vegansofcolor.wordpress.com/tag/angela-davis/.

Vermilya, Jenny R. "Contesting Horses: Borders and Shifting Social Meanings in Veterinary Medical Education." *Society and Animals* 20, no. 2 (2012): 123–137.

Viva! Health. "Britain's Hardest Working Mums." 2014. http://www.white-lies.org.uk/mothers.

Vosko, Leah F., ed. *Precarious Employment: Understanding Labour Market Insecurity in Canada.* Montréal and Kingston: McGill-Queens University Press, 2006.

———. *Temporary Work: The Rise of a Precarious Employment Relationship.* Toronto: University of Toronto Press, 2000.

Vosko, Leah F., Martha MacDonald, and Iain Campbell, eds. *Gender and the Contours of Precarious Employment.* New York: Routledge, 2009.

Wade, J. F., ed. *Proceedings of the 7th Colloquium on Working Equids.* Norwich: World Horse Welfare, 2014.

Walters, Kerry S., and Lisa Portmess. *Ethical Vegetarianism: From Pythagoras to Peter Singer.* Albany: SUNY Press, 1999.

Warkentin, Traci. "Whale Agency: Affordances and Acts of Resistance in Captive Environments." In *Animals and Agency: An Interdisciplinary Framework,* edited by Sarah E. McFarland and Ryan Hediger, 23–44. Leiden: Brill, 2009.

Waring, Marilyn. *If Women Counted: A New Feminist Economics.* Toronto: Harper Collins Canada, 1990.

Warren, Cat. *What the Dog Knows: The Science and Wonder of Working Dogs.* New York: Simon & Schuster, 2013.

Weis, Tony. *The Ecological Hoofprint: The Global Burden of Industrial Livestock.* London: Zed Books, 2013.

———. *The Global Food Economy: The Battle for the Future of Farming.* Halifax/London: Fernwood Press and Zed Books, 2007.

Weisberg, Zipporah. "Animal Assisted Intervention and Animal Citizenship." Unpublished paper, 2014.

———. "The Broken Promises of Monsters: Haraway, Animals and the Humanist Legacy." *Journal for Critical Animal Studies* 7, no. 2 (2009): 21–61.

———. "The Trouble with Posthumanism: Bacteria Are People, Too." In *Critical Animal Studies: Thinking the Unthinkable,* edited by John Sorenson, 93–116. Toronto: Canadian Scholars' Press, 2014.

Westlund, Stephanie. *Field Exercises: How Veterans Are Healing Themselves Through Farming and Outdoor Activities.* Gabriola Island: New Society Publishers, 2014.

Westoll, Andrew. *The Chimps of Fauna Sanctuary: A True Story of Resilience and Recovery.* Toronto: HarperCollins, 2011.

Whiten, Andrew, Jane Goodall, William C. McGrew, Toshisada Nishida, Vernon Reynolds, Yukimaru Sugiyama, Caroline E. G. Tutin, Richard W. Wrangham, and Christophe Boesch. "Cultures in Chimpanzees." *Nature* 399, no. 6737 (1999): 682–685.

Wilde, Lawrence. "'The Creatures, Too, Must Become Free': Marx and the Animal/Human Distinction." *Capital & Class* 24, no. 3 (2000): 37–53.

Wilkie, Rhoda M. *Livestock/Deadstock: Working with Farm Animals from Birth to Slaughter.* Philadelphia: Temple University Press, 2010.

Williams, Jane, Katherine Smith, and Fernando DaMata. "Risk Factors Associated with Horse-falls in UK Class 1 Steeplechases: 1999–2011." *International Journal of Performance Analysis in Sport* 14, no. 1 (2014): 148–152.

Williams, Raymond. *The Country and the City.* Oxford: Oxford University Press, 1975.

Woldehanna, S., and Zimicki, S. "An Expanded One Health Model: Integrating Social Science and One Health to Inform Study of the Human-Animal Interface." *Social Science & Medicine* 129 (March 2015): 87–95.

Wolkowitz, Carol. *Bodies at Work.* London: Sage, 2006.

World Health Organization. 2010. "The FAO-OIE-WHO Collaboration: Sharing Responsibilities and Coordinating Global Activities to Address Health Risks at the Animal-Human-Ecosystems Interfaces." http://www.who.int/foodsafety/zoonoses/final_concept_note_Hanoi.pdf?ua=1.

World Animal Protection. "Protecting Working Animals." http://www.worldanimalprotection.ca/ourwork/animalsincommunities/workingani mals/Protecting-working-animals-in-the-west-bank.aspx.

World Society for the Protection of Animals. *Curb the Cruelty: Canada's Farm Animal Transport System in Need of Repair.* Toronto: World Society for the Protection of Animals, 2010.

Zamir, Tzachi. "The Moral Basis of Animal-Assisted Therapy." *Society and Animals* 14, no. 2 (2006): 179–199.

Zeder, Melinda A. "Domestication and Early Agriculture in the Mediterranean Basin: Origins, Diffusion, and Impact." *Proceedings of the National Academy of Sciences* 105, no. 33 (2008): 11597–11604.

Index

CPSIA information can be obtained
at www.ICGtesting.com
Printed in the USA
LVHW081708201019
634742LV00006B/10/P